T0137120

Intelligent Systems Reference Library

Volume 103

Series editors

Janusz Kacprzyk, Polish Academy of Sciences, Warsaw, Poland
e-mail: kacprzyk@ibspan.waw.pl

Lakhmi C. Jain, Bournemouth University, Fern Barrow, Poole, UK, and
University of Canberra, Canberra, Australia
e-mail: jainlc2002@yahoo.co.uk

About this Series

The aim of this series is to publish a Reference Library, including novel advances
and developments in all aspects of Intelligent Systems in an easily accessible and
well structured form. The series includes reference works, handbooks, compendia,
textbooks, well-structured monographs, dictionaries, and encyclopedias. It contains
well integrated knowledge and current information in the field of Intelligent Systems. The series covers the theory, applications, and design methods of Intelligent
Systems. Virtually all disciplines such as engineering, computer science, avionics,
business, e-commerce, environment, healthcare, physics and life science are
included.

More information about this series at http://www.springer.com/series/8578

Alina Bărbulescu

Studies on Time
Series Applications in
Environmental Sciences

 Springer

Alina Bărbulescu
Ovidius University of Constanta
Constanta
Romania

ISSN 1868-4394   ISSN 1868-4408 (electronic)
Intelligent Systems Reference Library
ISBN 978-3-319-80809-3   ISBN 978-3-319-30436-6 (eBook)
DOI 10.1007/978-3-319-30436-6

Printed on acid-free paper

This Springer imprint is published by Springer Nature
The registered company is Springer International Publishing AG Switzerland

To Prof. Dr. Eng. Radu Dobrot—
*To **the One**,*
that made me laugh in the most difficult moments,
that encouraged and supported me—
with all my gratitude and love

Preface

Modeling and forecasting hydro-meteorological time series are of great interest due to their practical applications and impact on the human life. Many methods have been successfully used for solving different types of problems in this area. Here, we try only to summarize some possible approaches and to present a part of our results concerning modeling particular hydrological time series, from a region less studied in Europe.

The book is divided into seven chapters. In the first one we introduce the data series. The next two chapters are mainly theoretical. They contain some tests used for checking different statistical hypotheses on univariate time series and their implementation in R software, as well as a very short overview on the modeling techniques applied in the next chapters.

The rest of this book contains a part of our results, obtained the last 7 years on modeling the precipitation series or applications of some methods proposed by other scientists for generation precipitation fields.

One chapter summarizes the results of modeling the pollutants' dissipation in the atmosphere, while another one summarizes that of the evolution of water quality of two lakes, which are known to be affected by the atmospheric pollution and human activities.

All models refer to series from Dobrogea, a region situated in the southeastern part of Romania, between the Black Sea and the Danube River, for which no systematic study of climatic evolution has been done till 2007.

Scientific research is usually a team work, so a part of the results presented here has been obtained in cooperation with Dr. Lucica Barbeş, Dr. Elena Băutu, Dr. Judicael Deguenon, Dr. Cristina Gherghina, Dr. Carmen Elena Maftei, Dr. Elena Pelican, Dr. Nicolae Popescu-Bodorin, and Dr. Dana Simian. Thanks for their cooperation.

All the gratitude goes to Prof. Dr. Eng. Radu Victor Drobot, my Ph.D. supervisor in Civil Engineering. Without his valuable suggestions on different research topics and his continuous encouragement, this book wouldn't exist.

Thanks to Dr. Lakhmi Jain for his invitation to write this book.

Last but not least, thanks to my family that supported me unconditionally.

Sharjah, UAE   Alina Bărbulescu
October 2015

Contents

Data Series

The study region is Dobrogea, which is situated in the southeastern part of Romania, between the Black Sea and the Danube River, between 27°15′05″ and 29°30′10″ Eastern longitude and 43°40′04″ and 45°25′03″ Northern latitude. It has a surface of 11,145 km^2 without the Danube Delta and the lake basin Razim-Sinoe [7] (Fig. 1).

Fig. 1 Dobrogea region, and the hydrological and hydro-meteorological stations. The main stations have roman letters attached

Its active surface is not uniform and it presents interesting geo-morphological characteristics. Since the presentation of these features is out the scope of this book, we invite the interested readers to refer to [4, 7].

The average annual temperature is 11 °C in the western part of the territory and is over 11 °C in its northeastern and northern parts. The mean annual temperature decreases from south to north, concomitantly with the altitude increasing and the augmentation of the continental influence inside region [5].

Different aspects related to the temperature variations and modeling of climate in this region can be found in [1–3, 6].

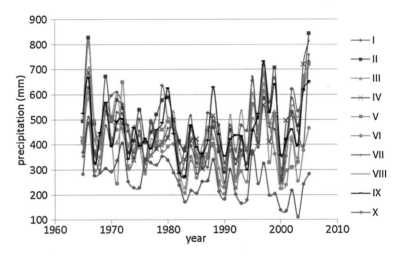

Fig. 2 Annual precipitation series recorded at the main stations

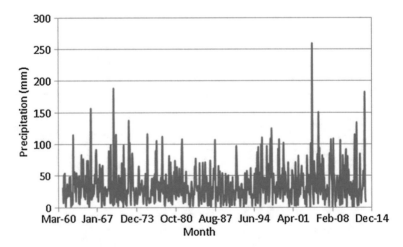

Fig. 3 Constanta monthly precipitation series (January 1961–December 2013)

The annual and monthly series studied here are recorded in the period January 1965–December 2005 at the ten main stations (that has a roman letter attached on the map on Fig. 1) and 41 secondary ones. In Fig. 2, we present the annual series of precipitation record in the period 1995–2005. For some series, as Constanta (Fig. 3) and Sulina, longer periods are used. All the series are complete, without gaps and reliable, collected by INHGA or from http://eca.knmi.nl/.

References

1. Bărbulescu, A.: Modeling temperature evolution. Case study, Romanian Reports in Physics, **68**(1), (2016) (to appear)
2. Bărbulescu, A., Băutu, E.: ARIMA and GEP models for climate variation, International Journal of Mathematics and Computation, **3**(J09), pp. 1–7 (2009)
3. Bărbulescu, A., Băutu, E.: Mathematical models of climate evolution in Dobrudja, Theoretical and Applied Climatology, **100**(1–2), pp. 29–44 (2010)
4. Bărbulescu, A., Deguenon, J., Teodorescu, D.: Study on Water Resources in the Black Sea Region, Nova Publishers, USA (2011)
5. Bărbulescu, A., Teodorescu, D., Deguenon, J.: Study on water resources in the Romanian Black Sea region. In Ryann, A.M., Perkins, N.J. (eds), The Black Sea: Dynamics, Ecology and Conservation, pp. 175–205. Nova Publishers, USA (2011)
6. Maftei, C., Bărbulescu, A.: Statistical analysis of climate evolution in Dobrudja region, World Congress on Engineering, Imperial College of London, England, July 02–04, 2008, Book series: Lecture Notes in Engineering and Computer sciences, vol. II, pp. 1082–1087 (2008)
7. Posea, G., Bogdan,O., Zăvoianu, I.: Geografia Romaniei, vol. **V**, Editura Academiei Române, Bucuresti, (2005) (in Romanian)

Chapter 1
Hypotheses Testing on Meteorological Time Series

Data analysis is an important step in time series modeling and forecasting. It requires: obtaining and preparing the dataset, exploratory analysis of data series, performing statistical tests and the results' interpretation, the former being of major importance for all the other stages of analysis. Dealing with hydro-meteorological time series requires attention to data acquisition (measurement methodology and frequency), the series length (at least 50 years, in the case of studies concerning the climate change) and their completeness [105].

In this chapter, we focus on testing some statistical hypothesis and methods of detecting the long range dependence property in hydro-meteorological time series.

In what follows we denote the observed data by $(x_i)_{i=\overline{1,n}}$. They are realizations of a time series process, denoted by $(X_i)_{i=\overline{1,n}}$.

1 Normality Tests

The Gaussian distribution is well-known, its properties being established for many years ago. It is a reference distribution to which one may report the experimental data to discover their properties. Testing the series normality is also important because many statistical methods rely on the hypothesis that the series are Gaussian.

The simplest way to check the null hypothesis (H_0: the series is normally distributed) against its alternative (H_1: the series is not normally distributed) is the use of graphical representations. One of them is the quantile–quantile plot (Q-Q plot) employed for deciding whether a univariate random sample comes from a given distribution G. The Q-Q plot is obtained by plotting the quantiles of the sample against the theoretical quantiles of G. If the sample comes from the specified distribution (in our case, the Gaussian one) then the points are close to a straight line.

MINITAB, SPSS, R have the option to draw the Q-Q plot. We mention that R is freeware software.

© Springer International Publishing Switzerland 2016
A. Bărbulescu, *Studies on Time Series Applications in Environmental Sciences*,
Intelligent Systems Reference Library 103, DOI 10.1007/978-3-319-30436-6_1

In the following we present the R code for obtaining the Q-Q plot of Constanta annual precipitation series (1961–2013).

```
data<- read.csv("D:\\Lucrari_2.12.14\\2015_Carte\\Cta_annual_1961_2013.csv", sep=",",
header=TRUE)
qqnorm(y); qqline(y, col = 2)
```

Looking to the chart (Fig. 1a) we could not reject H_0. We couldn't also decide the opposite since the points from the upper part of the Q-Q plot are not very close to the line.

The same decision could be taken, looking to the histogram of the series (Fig. 1b), which is a two-dimensional representation of the observed data against their frequency.

The R-function used to create the histogram is: hist(y, right=FALSE).

The decision to reject the normality hypothesis can be taken very easy for Constanta monthly series (1961–2013) because the plots significantly deviate from the straight line and the histogram is right-skewed (Fig. 2).

Since the decision on data normality is still difficult for Constanta annual series, other methods, based on statistical tests, are more appropriate. Among the tests used for this aim we mention: Kolmogorov–Smirnov [78, 111], Jarque–Bera [72], Shapiro–Wilk [110], Lilliefors [82] (presented in detail in [7]), Anderson–Darling [2] and Cramer von Mises [113].

The Anderson–Darling test is used to determine if a dataset comes from a specified distribution (the normal one, in our case), so it is considered to be also a goodness of fit test. It compares the fit of an observed cumulative distribution function to the expected cumulative distribution function.

Considering the sample data $\{x_i\}_{i=\overline{1,n}}$, and $\{x_{(i)}\}_{i=\overline{1,n}}$ its values increasingly ordered, the Anderson–Darling statistic is defined by:

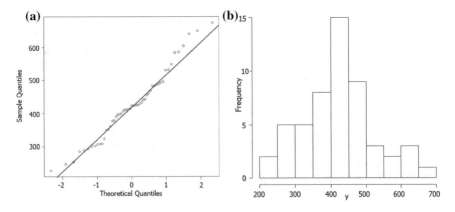

Fig. 1 a Q-Q plot and **b** histogram of Constanta annual precipitation series (1961–2013)

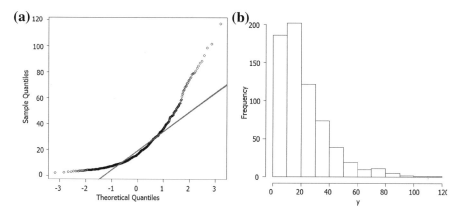

Fig. 2 a Q-Q plot and **b** histogram of Constanta monthly precipitation series (1961–2013)

$$A^2 = -n - \frac{1}{n}\sum_{i=1}^{n}(2i-1)[\ln F(x_{(i)}) + \ln(1 - F(x_{(n-i+1)}))], \qquad (1)$$

where F is the cumulative distribution function of the specified distribution.

If Anderson–Darling is used as normality test, (1) becomes:

$$A^2 = -n - \frac{1}{n}\sum_{i=1}^{n}(2i-1)[\ln p_{(i)} + \ln(1 - p_{(n-i+1)}],$$

where:

$$p_{(i)} = \Phi([x_{(i)} - \bar{x}]/s), \qquad (2)$$

Φ being the cumulative distribution function of the standard normal distribution, $\bar{x}$, the mean and s—the standard deviation of $\{x_i\}_{i=\overline{1,n}}$.

The Cramer-von Mises test for the composite hypothesis of normality is also based on the cumulative distribution function. The test statistic is:

$$W = \frac{1}{12n} + \sum_{i=1}^{n}\left(p_{(i)} - \frac{2i-1}{2n}\right),$$

where $p_{(i)}$ is given in (2).

The Pearson chi-square test for normality is applied to binned data so that the value of the statistic of the test depends on how the data was binned. Firstly, the data are standardized by subtracting the sample mean and dividing each value by the sample standard deviation. Then, the number of bins (k) is chosen by a formula (there is no optimal formula for this choice!), as, for example, $k = 1 + \log_2 n$, where n is the sample data.

The test statistic is:

$$\chi^2 = \sum_{i=1}^{k} (o_i - e_i)^2 / e_i,$$

where o_i is the observed frequency of the ith bin, e_i is the expected frequency of the ith bin, calculated by:

$$e_i = F(x_2) - F(x_1),$$

F is the cumulative distribution function of the distribution being tested, and x_1, x_2 are the limits of the ith bin.

When the mean and variance are known, the test statistic is asymptotically χ^2 distributed with $k - 1$ degrees of freedom. Therefore, the null hypothesis is rejected at a significance level α if the test statistic is greater than the quartile $\chi^2_{1-\alpha}(k - 1)$.

Simulation studies showed that the normality tests have different powers. The relative powers of the discrete statistics tests of Kolmogorov–Smirnov, Cramér-von Mises, Anderson–Darling and Watson and of two test statistics for nominal data (chi-square and Kolmogorov–Smirnov) for an uniform null distribution against a selection of fully specified alternative distributions has been studied in [112]. The results show that the Pearson's chi-square [96] and the nominal Kolmogorov–Smirnov are more powerful for the studied triangular, sharp and bimodal alternative distributions.

In the same idea, to compare the power of the Shapiro–Wilk, Kolmogorov–Smirnov, Lilliefors and Anderson–Darling tests, the Monte Carlo procedure was employed in [102]. The results proved that the least powerful test is the Kolmogorov–Smirnov one, and the most powerful one, Shapiro–Wilk, followed by Anderson–Darling. However, the Shapiro–Wilk test is biased if the sample size and the significance level are small (that is the power could be less than α). Due to its inferior power by comparison to the other tests, it is not advisable to use the Pearson chi-square test for checking the composite hypothesis of normality [62, 91].

In the following, the statistical tests are performed at a significance level of 0.05, if another level is not specified.

To perform normality tests using the R software, one has to install the packages **fBasics** [55] and **nortest** [62] and to write the following commands:

```
library(fBasics)
ksnormTest(y, title = NULL, description = NULL)  # performs the Kolmogorov –Smirnov
test
jarqueberaTest(y, title = NULL, description = NULL)  # performs the Jarque - Bera test
shapiroTest(y, title = NULL, description = NULL)  # performs the Shapiro - Wilk test
library(nortest)
lillieTest(y, title = NULL, description = NULL)  # performs the Lilliefors test
ad.test(y)  # performs the Anderson - Darling test
cvmTest(y, title = NULL, description = NULL)  # performs the Cramer-von  Mises test
pearson.test(y, n.classes = ceiling(2 * (n^(2/5))), adjust = TRUE)
# performs the Pearson chi - square test, where:
    # n is the sample volume,
    # y is the vector containing the data values,
    # n.classes is the number of classes,
    # adjust - logical; if TRUE (default), the p-value is computed from a chi-square
# distribution with n.classes-3 degrees of freedom, otherwise from a chi-square distribution
# with n.classes-1 degrees of freedom.
```

Remarks 1. y is a numeric vector containing a minimum of 3 and maximum of 5000 values. In our case, it contains 53, respectively 636 values.

2. For all tests, but Jarque–Bera, the R software provides the values of the test statistics and the p-values. For the Jarque–Bera test it returns $\chi^2(2)$ and the asymptotic p-value. For Cramer-von Mises and Anderson–Darling tests, the p-values are computed respectively from the statistics $Z = W(1 + 0.5/n)$ and $Z = A(1 + 0.75/n + 2.25/n^2)$ [113]. The reader may also refer to [55, 62].

3. In the case of discrete distributions, the Kolmogorov–Smirnov and Cramer-von Misses tests can be performed in R using the functions ks.test() and cvm.test() from the package **dgof** [3]:

- ks.test() function supports one sample test for discrete null distributions. The second argument, y, can be an empirical cumulative distribution function (an R function with class "ecdf") or an object of class "stepfun" that specify a discrete distribution. For example:

```
library(dgof)
dgof::ks.test(c(1, 3), ecdf(c(2, 5)))
```

will return:

```
One-sample Kolmogorov-Smirnov test
data:  c(1, 3)
D = 0, p-value = 1
alternative hypothesis: two-sided
```

In the following, instead of writing "will return" before the results returned by R, we shall list only the results, after a white row and a Tab.

- The first two arguments of the function cvm.test() are the same as for ks.test(); the third one, type, specifies the variant of the Cramér–von Mises test used: W2 —is the default, U2—for cyclical data, A2—is the Anderson–Darling alternative.

We exemplify the application of this test on a sample of size 30, generated from the discrete uniform distribution on the natural numbers between 1 and 20:

```
x <- sample(1:20, 30, replace = TRUE)
cvm.test(x, ecdf(1:20))


        Cramer-von Mises - W2
        data:  x
        W2 = 0.0358, p-value = 0.947
        alternative hypothesis: Two.sided

cvm.test(x, ecdf(1:20), type= "A2")


        Cramer-von Mises - A2
        data:  x
        A2 = 0.2422, p-value = 0.9441
        alternative hypothesis: Two.sided

cvm.test(x, ecdf(1:20), type= "U2")


        Cramer-von Mises - U2
        data:  x
        U2 = 0.0275, p-value = 0.9386
        alternative hypothesis: Two.sided
```

The null hypothesis is rejected if the p-value is less than the significance level.

The results of some of the mentioned tests, applied to our data are given in Table 1. The normality hypothesis couldn't be rejected for the annual series, but it was rejected for the monthly one.

Table 1 Results of normality tests for Constanta precipitation series

Data		Jarque–Bera	Shapiro-Wilk	Lilliefors	Anderson-Darling	Cramer-von Mises
Annual	Stat.	1.3209	0.9722	0.0858	0.4672	0.073
	p-val	0.5166	0.2508	0.4284	0.2413	0.2503
Monthly	Stat.	765.6462	0.8324	0.1395	29.5533	0.998
	p-val	<2.2e−16	2.2e−16	2.2e−16	2.2e−16	1

Minitab offers the possibility of choosing among three normality tests: Anderson–Darling, Kolmogorov–Smirnov (but in fact, it is the Lilliefors one) and Ryan–Joiner (analogous to Anderson–Darling). SPSS also offers two alternatives: the Lilliefors and Shapiro–Wilk tests.

2 Homoskedasticity Tests

A sequence of random variables is said to be heteroskedastic if there are subsequences whose variances differ from the others. The heteroskedasticity is the opposite of homoscedasticity.

Many statistical tests have been proposed [10, 16, 48, 93] for testing the homoscedasticity in the hypothesis of data normality, or in regression models [14, 16, 44, 45, 122] etc. The null and alternative hypotheses in these tests are respectively:

$$H_0: \sigma_1^2 = \sigma_2^2 = \cdots = \sigma_k^2,$$
$$H_1: \sigma_i^2 \neq \sigma_j^2, \text{ for at least one pair } (i,j),$$

where k is the number of groups and σ_i^2 is the variance of the ith group.

The Bartlett test is the most used in applications that don't involve different types of regressions [10]. Since it is sensitive to the data departures from normality, other tests have been developed; among them, the Levene one [80], whose statistics is [7]:

$$W = \frac{n-k}{k-1} \cdot \frac{\sum_{i=1}^{k} n_i (\overline{Z_{i\cdot}} - \overline{Z_{\cdot\cdot}})^2}{\sum_{i=1}^{k} \sum_{j=1}^{n_i} (Z_{ij} - \overline{Z_{i\cdot}})^2},$$

where n is the total number of observations, n_i is the number of observations in the group i, Z_{ij} is the absolute value of the deviation of a value (X_{ij}) in a group from the group's mean $(\overline{X_{i\cdot}})$, $\overline{Z_{i\cdot}}$ is the average of the values Z_{ij} in the group i, $\overline{Z_{\cdot\cdot}}$ is the overall mean of Z_{ij}.

The null hypothesis is rejected at the significance level α if the p-value is less than α.

In the robust version of the Levene test, proposed by Brown–Forsythe [16], the group mean is replaced the group median.

Analyzing the sensitivity of these tests to the samples' lengths, O'Brien [95] introduced a correction factor. Hines [51] also proposed a method for improving the Brown–Forsythe test, which takes into account the fact that the possible linear relationships (structural zeroes) of the investigated data were not previously investigated. Noguchi and Gel [94] introduced a new correction factor based on a combination of [51] and [95].

Levene's test and its versions are implemented in R **lawstat** package [60].

```
library(lawstat) # loads the library lawstat
data<-read.csv("D:\\Lucrari_2.12.14\\2015_Carte\\Cta_lunar_1961_2013_grupuri.csv",
sep= "," , header=TRUE) # loads the file containing the monthly data
x<-data[,1]
y<-data[,2]
levene.test(x, y, location="mean") # performs Levene's test
```

    data: x
    Test Statistic = 3.0043, p-value = 0.05028

```
levene.test(x, y, location="median", correction.method="none") # performs the Brown –
# Forsythe test
```

    data: x
    Test Statistic = 2.314, p-value = 0.0997

```
levene.test(x, y, location="mean", correction.method="correction.factor") # performs
the Levene test with the correction factor [95].
```

    data: x
    Test Statistic = 3.0135, p-value = 0.04982

```
levene.test(x, y, location="median", correction.method="zero.correction") # performs the
# Brown-Forsythe test with modified structural zero removal method and correction factor
#[95].
```

    data: x
    Test Statistic = 2.3308, p-value = 0.09806.

Hartley's F_{max} [48] is a test for homogeneity of variances, used if all the samples have the same size and the data in each group are normally distributed. The F_{max} statistics is computed as the ratio of the largest and smallest variances of the groups. Tables for this test are due to David [27] and Nelson [92].

Cochran's test [23] was developed for testing the homogeneity of groups' variances against the hypothesis that the highest variance is different from the others if all the k groups contain the same number of data. The test statistic is:

$$Q = \frac{\max\limits_{1 \le i \le k} s_i^2}{\sum_{i=1}^{k} s_i^2},$$

where s_i^2 the variance of the ith group.

The homogeneity hypothesis is rejected at the significance level, α, if $Q > Q_{\alpha;k,n-1}$, where $Q_{\alpha;k,n-1}$ is the critical value from the tables of Cochran test, function of the sample volume, n, the number of groups, k, and the significance level [7].

Other tests have been designed for testing special types of heteroskedasticity as: the Breush–Pagan test [14], that assumes the heteroskedasticity as a result of a linear combination of independent variables or the White test [122], which is based on the regression of the squared residual from a given model onto the initial regressors, their squared product and cross-products.

The Breusch–Pagan and White tests are implemented in R in **lmtest** package [61], and **het.test** package [59], respectively.

3 Autocorrelation Tests

The autocorrelation (or serial correlation) refers to the correlation between the time series values with its previous/future values.

If (X_t) is a time series and $\gamma(h) = Cov(X_t, X_{t+h})$ is the *autocovariance function* of (X_t) at lag h $(h \in \mathbf{N}^*)$, then *the autocorrelation function of (X_t) at lag h is* defined by $\rho(h) = \gamma(h)/\gamma(0), h \in \mathbf{N}^*$.

The most used estimator of the autocorrelation function is *the empirical autocorrelation function* (ACF), defined by:

$$\hat{\rho}(h) = \frac{\sum_{t=1}^{n-h} (x_t - \bar{x})(x_{t+h} - \bar{x})}{\sum_{t=1}^{n} (x_t - \bar{x})^2},$$

where $(x_i)_{i=\overline{1,n}}$ are observed values of (X_t) and $\bar{x}$ is the average of $(x_i)_{i=\overline{1,n}}$.

The chart of ACF is called *correlogram*.

For taking the decision about the existence of the autocorrelation in a time series, the confidence interval at a selected confidence level (usually 0.95 or 0.99) is also computed, and the correlogram is analyzed. If all the values of the autocorrelation function are inside the empirical confidence interval, one can reject the autocorrelation hypothesis.

The Durbin–Watson test [33–35] is widely used for checking the null hypothesis of correlation's absence against the first order autocorrelation of residuals in the regression analysis. It is based on the hypothesis that the errors in the regression model are generated by an AR(1) process.

The test is inconclusive in the interval (d_l, d_u), whose limits, d_l, d_u, are specified in the tables of the test, at the specified significance level. Also, it cannot be used for testing the residuals' autocorrelation when there are lagged endogenous variables among the exogenous variables. To overcome this drawback, Durbin [32] introduced the *h*-test, whose statistics is:

$$h = \left(1 - \frac{d}{2}\right) \sqrt{\frac{n}{1 - n\sigma^2}},$$

where σ^2 is the estimated variance of the coefficient corresponding to the lagged dependent variable, β, n is the sample size and

$$d = \sum_{t=2}^{n} (x_t - x_{t-1})^2 \bigg/ \sum_{t=1}^{n} x_t^2$$

is the Durbin–Watson statistics.

The Durbin–Watson test is implemented in **lmtest** [61] package in R.

Other procedures, that will not be discussed here, as the Cochrane–Orcutt test [24], the balisage method of Hildreth–Lu [87] or generalized differencing, have also been proposed for dealing with serial correlation of errors in regression models.

The correlogram analysis is a good option for searching for a linear dependent structure in a time series; otherwise, different autocorrelation tests must be used. In the following, we shortly present the so-called *portmanteau tests*, designed for testing the null hypothesis that the residuals in a model form a white noise.

Box and Pierce [13] portmanteau test was built for testing the hypothesis that the residual in an ARMA(p, q) model is a white noise. It is based on the statistics:

$$Q_{BP} = n \sum_{k=1}^{h} \hat{\rho}_k^2,$$

where n is the sample size, $\hat{\rho}_k$ is the sample autocorrelation of order k, and h is the number of lags.

Since simulation studies showed low performances of this statistics, different improvements have been proposed. One of them is that of Ljung and Box [85], whose statistics is defined by:

$$Q_{LB} = n(n+2) \sum_{k=1}^{h} \frac{\hat{\rho}_k^2}{n-k}.$$

If $Q_{LB} > \chi^2_{1-\alpha, h-p-q}$, then the null hypothesis of residuals' independence is rejected, where $\chi^2_{1-\alpha, h-p-q}$ is the α—quantile of the chi—square distribution with $h - p - q$ degrees of freedom.

This test can be also used to test the hypothesis of independence for any time series, replacing $h - p - q$ by h.

Other portmanteau tests have been introduced by:

- Monti [90], whose statistics is:

$$Q_M = n(n+2) \sum_{k=1}^{h} \frac{\hat{\tau}_k^2}{n-k},$$

where $\hat{\tau}_k^2, 1 \leq k \leq h$ is the value of the partial autocorrelation function of residuals (see the next chapter for definition);

- Li and McLeod [81], whose statistics is:

$$Q_{ML} = n(n+2) \sum_{k=1}^{h} \frac{\hat{\rho}_{aa}^2(k)}{n-k},$$

where $\hat{u}_t$ are the estimated residuals in an ARMA model,

$$\hat{\rho}_{aa}(k) = \frac{\sum_{t=1}^{n-h} (\hat{u}_t^2 - \bar{\mu}^2)(\hat{u}_{t+h}^2 - \bar{\mu}^2)}{\sum_{t=1}^{n} (\hat{u}_t^2 - \bar{\mu}^2)^2},$$

and $\bar{\mu}$ is the mean of $\hat{u}_t^2$.

Q_M and Q_{ML} follow asymptotic chi-square distributions with $m - p - q$ and m degrees of freedom, respectively, in the hypothesis that the fitted ARMA model is correct. For more details on these tests, see [4].

The Box–Pierce, Ljung–Box and Li and McLeod tests are implemented in **portes** package [64] in R.

In the following, we illustrate the use of a part of these tests on the monthly precipitation data and the residual from the ARMA(1,1) model.

Firstly the package must be installed, via 'Install Package'. Then the following instructions must be written for performing the Box–Pierce test on the data series:

```
library(portes) # loads the library 'portes'
data<-read.csv("D:\\Lucrari_2.12.14\\2015_Carte\\Cta_annual_1961_2013.csv", sep=",",
header=TRUE) # loads the file containing the monthly data
x<-data
y<-ts(x) # defines the time series y using the input data
BoxPierce(y) # performs the Box - Pierce test on the data
```

Lags	Statistic	df	p - value
5	4.205184	5	0.5202686
10	8.874683	10	0.5440365
15	11.503326	15	0.7161672
20	15.415612	20	0.7521529
25	17.554509	25	0.8607075
30	17.971275	30	0.9589969

```
LjungBox(y) # performs the Ljung - Box test on the data
```

Lags	Statistic	df	p - value
5	4.67418	5	0.4569221
10	10.34766	10	0.4105394

15	13.93453	15	0.5305002
20	20.21133	20	0.4447819
25	24.04476	25	0.5167882
30	24.96289	30	0.7268334

fit <- arima(x, c(1, 0, 1))) # *fits and ARMA(1,1) model to the data*
lags <- c(5, 10, 15, 20,35, 30) # *defines the lags*
res <- resid(fit) # *determines the residual in the ARMA(1,1) model*
BoxPierce(res, lags, order = 2) # *applies the Box-Pierce test on the residual of ARMA(1,1)*
model; specification of the order is required for computing the degrees of freedom of
asymptotic chi - square distribution. Generally, it is equal to the number of estimated
parameters in the model.

Lags	Statistic	df	p - value
5	4.385739	3	0.2227114
10	7.820398	8	0.4512081
15	10.363177	13	0.6640005
20	13.708626	18	0.7478784
25	14.973888	23	0.8955670
30	15.408471	28	0.9737939

LjungBox(res, lags, order = 2) # *performs the Ljung - Box test on the residual of*
ARMA(1,1) model

Lags	Statistic	df	p - value
5	4.997150	3	0.1720060
10	9.180725	8	0.3272805
15	12. 647649	13	0.4753816
20	17.974094	18	0.4573605
25	20.271432	23	0.6254872
30	21.236837	28	0.8153370

LiMcLeod(res, lags, order = 0, SquaredQ=TRUE) # *applies the Li - Mc Leod test on the*
residual of ARMA(1,1) model. Since the study object is issued from an ARMA model, the
order' specification is not necessary, being automatically computed. If SquaredQ = TRUE,
the statistics value in the Li - McLeod test is computed by squared values because it checks
for ARCH effects. If SquaredQ = FALSE, the residual are used for the computation of the
test - statistics.

Lags	Statistic	df	p - value
5	4.300665	5	0.5069880
10	9.692352	10	0.4678862
15	14.995210	15	0.4517623
20	19.148617	20	0.5121855
25	24.844189	25	0.4711326
30	30.578967	30	0.4363097

The tests' results reject the autocorrelation hypothesis of the data or residual in the fitted model.

Other statistical tests for serial independence are based on ranks, on empirical copulae, on divergent measures, spectral theory, information theory, etc. For the classification of these methods, the reader may refer to [29].

Practical needs of working with many random processes or with multivariate data series lead to the development of statistical independence tests based on multivariate extensions of Spearman's ρ-test [39, 42, 119]. They also imply the development of other nonparametric [6, 12, 28, 100, 103], parametric [114] or symbolic [89] techniques for testing the statistical independence.

In one of the following chapters we shall use the Szekely–Rizzo-Bakirov test (SRB) for independence [115]. We shortly present it in the following.

For any two multivariate random variables (random vectors) $X \in \mathrm{R}^p$ and $Y \in \mathrm{R}^q$ with finite expectations, the *distance covariance* $v^2(X, Y)$ and the *distance correlation* $R^2(X, Y)$ are defined in [115] as:

$$v^2(X, Y) = \|f_{XY} - f_X f_Y\|^2,$$

$$R^2(X, Y) = \begin{cases} 0; & v^2(X)v^2(Y) = 0 \\ \frac{v^2(X,Y)}{\sqrt{v^2(X)v^2(Y)}}; & v^2(X)v^2(Y) \neq 0 \end{cases}$$

where f_X, f_Y, f_{XY} are respectively the characteristic functions of X, Y and the joint characteristic function of (X, Y).

X and Y are statistical independent if:

$$f_{XY} = f_X f_Y, \quad v^2(X, Y) = 0 \quad \text{and} \quad R^2(X, Y) = 0,$$

Practically, when the analytical expressions of the two distributions are not known, the only way of knowing X and Y is by recording their samples of the same length n, registered as matrices $\mathbf{X}_n \in \mathrm{R}^{n \times p}$, $\mathbf{Y}_n \in \mathrm{R}^{n \times q}$. When n is big, the sample matrices $(\mathbf{X}_n, \mathbf{Y}_n) = \{(X_k, Y_k) | k \in \overline{1, n}\} \in \mathrm{R}^{n \times p} \times \mathrm{R}^{n \times q}$ becomes representative for the variables (X, Y) that produce them.

Therefore, the *empirical distance covariance* (EDCov, $v_n^2(X, Y)$), and the *empirical distance correlation* (EDCor, $R_n^2(\mathbf{X}_n, \mathbf{Y}_n)$) are introduced [115] using only the experimental data:

$$R_n^2(\mathbf{X}_n, \mathbf{Y}_n) = \begin{cases} 0, & v_n^2(\mathbf{X}_n)v_n^2(\mathbf{Y}_n) = 0 \\ \frac{v_n^2(\mathbf{X}_n,\mathbf{Y}_n)}{\sqrt{v_n^2(\mathbf{X}_n)v_n^2(\mathbf{Y}_n)}}, & v_n^2(\mathbf{X}_n)v_n^2(\mathbf{Y}_n) \neq 0 \end{cases},$$

respectively

$$v_n^2(\mathbf{X}_n, \mathbf{Y}_n) = \frac{1}{n^2} \sum_{k,l=1}^{n} A_{kl} B_{kl},$$

where

$$v_n^2(\mathbf{X}_n) = v_n^2(\mathbf{X}_n, \mathbf{X}_n), v_n^2(\mathbf{Y}_n) = v_n^2(\mathbf{Y}_n, \mathbf{Y}_n),$$

$$A_{kl} = a_{kl} - \bar{a}_{k\cdot} - \bar{a}_{\cdot l} + a_{\cdot\cdot}, B_{kl} = b_{kl} - \bar{b}_{k\cdot} - \bar{b}_{\cdot l} + b_{\cdot\cdot},$$

$$a_{kl} = \|X_k - X_l\|_p, b_{kl} = \|Y_k - Y_l\|_q,$$

$$\bar{a}_{k\cdot} = \frac{1}{n} \sum_{l=1}^{n} a_{kl}, \bar{b}_{k\cdot} = \frac{1}{n} \sum_{l=1}^{n} b_{kl}, \bar{a}_{\cdot l} = \frac{1}{n} \sum_{k=1}^{n} a_{kl}, \bar{b}_{\cdot l} = \frac{1}{n} \sum_{k=1}^{n} b_{kl},$$

$$a_{\cdot\cdot} = \frac{1}{n^2} \sum_{k,l=1}^{n} a_{kl}, b_{\cdot\cdot} = \frac{1}{n^2} \sum_{k,l=1}^{n} b_{kl}.$$

It was proved [115] that:

$$v_n^2(\mathbf{X}_n, \mathbf{Y}_n) = S_1^n + S_2^n - 2S_3^n = \left\| f_{XY}^n - f_X^n f_Y^n \right\|^2,$$

where f_X^n, f_Y^n, f_{XY}^n are respectively the characteristic functions of $\mathbf{X}_n$, $\mathbf{Y}_n$ and the joint characteristic function of $(\mathbf{X}_n, \mathbf{Y}_n)$. S_1^n, S_2^n, S_3^n are computed in terms of L_p, L_q norms related to $\mathbf{X}_n$ and $\mathbf{Y}_n$.

Also:

$$\lim_{n \to \infty} v_n^2(\mathbf{X}_n, \mathbf{Y}_n) = v^2(X, Y) \, a.s./a.e.;$$

$$\lim_{n \to \infty} R_n^2(\mathbf{X}_n, \mathbf{Y}_n) = R^2(X, Y) \, a.s./a.e.,$$

so the statistical independence of X and Y is proved if $\lim_{n \to \infty} v_n^2(\mathbf{X}_n, \mathbf{Y}_n) = 0 \, a.e.$

Practically, applying SRB means the computation of $v_n^2(\mathbf{X}_n, \mathbf{Y}_n)$, followed by testing for EDCor convergence.

The implementation of the procedure described above has been done by Maria L. Rizzo and Gabor J. Szekely, in the R package **energy** [58] . We present here the results of its application to two matrices containing five annual precipitation series.

```
data<- read.csv("D:\\Lucrari_2.12.14\\2015_Carte\\Anuale_carte.csv", sep=",",
header=TRUE) # reads the data containing ten annual precipitation series
library(energy) # loads the library
x<- data[1:41,2:6] # defines the first matrix, containing the first data series
y<-data[1:41,7:11] # defines the second matrix, containing the last five data series
DCOR(x, y) # computes the empirical distance covariance EDCov, the empirical distance
# correlation variance, distance variance of x and distance variance of x
```

$dCov
[1] 104.9466

$dCor
[1] 0.8806615

$dVarX
[1] 135.1349

$dVarY
[1] 105.0874

4 Outliers' Detection

An outlier (aberrant value) is a value that appears to deviate markedly from other members of the sample in which it occurs [8], or that differs so much from the other observations such as there are suspicions that it was produced by a different mechanism [49].

There are different methods for outliers' detection, which can be classified, for example, as: statistical-based approaches (parametric—Gaussian model-based, regression model-based—or nonparametric—histogram-based, kernel-based), nearest neighbor-based, clustering-based (SVM-based, Bayesian network-based), classification-based and spectral decomposition-based (Principal Component Analysis-based) approaches [20, 123].

In outlier analysis, different domains of data require specific detection techniques. Temporal outlier analysis studies the aberrant values of the data across time. Given a time series, one can find particular elements as outliers (point outliers, generally referred as outliers), or subsequence outliers. In our study, we shall discuss only the first type of outliers.

The most popular univariate parametric methods rely on the assumption that data distribution is known, and the process that produces them is independent, identically distributed or Gaussian. Overviews of techniques for outliers' detection have been provided by many authors [1, 20, 52]. Here we shortly present some statistical methods.

Dixon's method [30], developed in 1950, is used for samples with small dimensions (up to 40) and is based on order statistics. Different ratios are defined for identification of the potential outliers, function of the number of presumed aberrant values. A value is considered to be an outlier when the corresponding statistics value is greater than the critical value of the test.

Grubbs' method [46] is used for testing the null hypothesis that there is no outlier in the data series $(x_i)_{i=\overline{1,n}}$, against the alternative that there is at least an outlier, based on the statistics:

$$G = \max_{i=\overline{1,n}} \frac{|x_i - \bar{x}|}{s},$$

where $\bar{x}$ is the sample mean and s is the standard deviation of the $(x_i)_{i=\overline{1,n}}$.

If at a chosen significance level, α, G is higher than the critical value for Grubbs' test, then the null hypothesis is rejected and the corresponding x_i can be accepted as an outlier.

Note that the test assumes data normality.

The Tietjen–Moore test is a generalized version of Grubbs' method. It is designed for the detection of multiple outliers in a univariate data series which is approximately normally distributed, when the suspected number of outliers, k, is correctly specified. More precisely, the hypothesis that no outlier exists in the data series is tested against the alternative that there exist exactly k outliers. After sorting the sample in ascending order, different test statistics are defined, for the k largest points, the k smallest points or for testing the outliers' existence in both tails. The values of the test statistics are in the interval [0, 1]. In the presence of outliers, the value is close to zero and in their absence, it is 1. The computation of the test critical region is done by simulation.

The main drawback of this test is that the number of outliers must be exactly specified. To surpass this inconvenience, *the generalized ESD test* (Extreme Studentized Deviate) [106] can be used. In this test, the hypothesis that there is no outlier in the data series is tested against that of the existence of up to r outliers. It is also supposed that the process generating the data series follows an approximately Gaussian distribution. After computing the statistics

$$R_i = \max_{i=\overline{1,n}} \frac{|x_i - \bar{x}|}{s},$$

the observation that maximizes $|x_i - \bar{x}|$ is removed and R_i is recomputed with the remained data. The process is repeated until r observations have been removed. Corresponding to each test statistics, the critical value is computed by:

$$\lambda_i = \frac{(n-i)t_{p,n-i-1}}{\sqrt{(n-i-1+t_{p,n-i-1}^2)(n-i+1)}}, \quad i = \overline{1,r},$$

where: $t_{p,n-i-1}$ is the $100p$ percentage point from the Student distribution with $n-i-1$ degrees of freedom and $p = 1 - \frac{\alpha}{2(n-i+1)}$, α being the significance level.

The number of outliers is defined by:

$$N = \max\{i \colon R_i > \lambda_i\}.$$

This method is better than the Grubbs' one because it adjusts the critical values function of the outliers' number taken into account.

When the process is not Gaussian, a box-plot is suggestive for the outliers' identification.

A box plot is a graphical representation of the data dispersion, which draws the quartiles (first, third and the median), together with two fences. Any observation situated outside these fences is considered a potential outlier.

Tests for outliers' detection are implemented in R, in the packages: **outliers** (for univariate data series) [63], **mvoutliers** (for multivariate data series), **tsoutliers** (for time series, based on ARIMA models), and **extremevalues**.

For performing the Grubbs and Dixon tests for outliers' detection and for drawing the box plot for Constanta annual series, the following sequence of code is written in R:

Fig. 3 Box plot of: **a** Constanta annual precipitation series (1965–2005); **b** Constanta annual precipitation series (1965–1994)

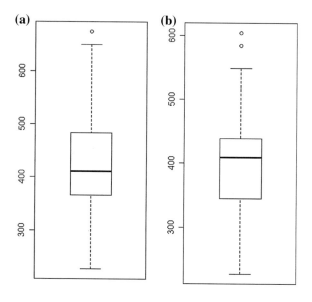

```
data<- read.csv("D:\\Lucrari_2.12.14\\2015_Carte\\Anuale_carte.csv", sep=",",
header=TRUE) # reads the data containing ten annual precipitation series
  x<- data [,4] # reads the data containing Constanta annual precipitation series for the
# period 1965-2005
  library(outliers) # loads the library
  grubbs.test(x, type=10) # performs Grubbs' test for the detection one outlier, statistically
# different from the other values. If type = 11, the test is used for checking if the lowest and
# highest values are outliers. If type = 20 the test is used to verify if data series contains
# two outliers on the same tail. In this case, the sample volume must be between 3 and 30.
```

Grubbs test for one outlier
data: x
G = 2.2971, U = 0.8648, p-value = 0.3694
alternative hypothesis: highest value 674.8 is an outlier

```
grubbs.test(x, type=11)
```

Grubbs test for two opposite outliers
data: x
G = 4.0857, U = 0.7879, p-value = 1
alternative hypothesis: 227 and 674.8 are outliers

```
y<-data [1:30,4] # reads the first 30 values in Constanta annual precipitation series
grubbs.test(y, type=20)
```

Grubbs test for two outliers
data: y
U = 0.7103, p-value = 0.3181
alternative hypothesis: highest values 548.7 , 584.3 are outliers

```
dixon.test(y, type = 0, opposite = FALSE, two.sided = TRUE) # performs Dixon's test on y
# type is a natural number that selects the test statistic, function of the sample volume. If
# it is zero, the selection is automatically performed. opposite is a parameter used for the
# selection of the extreme value (maximum or minimum).
```

Dixon test for outliers
data: y
Q = 0.157, p-value = 0.8301
alternative hypothesis: highest value 584.3 is an outlier

```
boxplot(x) # draws the box plot for the series x (Fig. 3a.)
boxplot(y) # draws the box plot for the series y (Fig. 3b)
```

For performing the Tietjen–Moore and ESD tests we used the codes from [66, 67], adapted to the series y from the previous example and to $k = 2$ outliers. The result of the first test is: the value of the test statistics is 0.7099646, and the critical value at 0.05 level of significance is 0.5477848. The result of the second test is:

No. Outliers	Test Stat.	Critical Val.
1	2.174088	2.908473
2	1.987073	2.892705

In both cases the test statistics are smaller than the critical values, so the hypothesis of outliers' presence can be rejected.

5 Change Points' Detection

It is known that meteorological time series are affected by more or less abrupt changing conditions in the environment. A breakpoint or a change point is defined to be the moment when the process generating the series changes.

The literature usually treats the changes in mean, but changes in variance, in frequency structure or in the system are also considered. We focus here on changes in the mean.

Solutions to this problem have been proposed in [11, 15, 21, 26]. The earliest studies considered the case of a sequence of independent identical and normally distributed variables for which the change point in the mean had to be detected [22]. Later, Bayesian [19, 71, 101] and non-Bayesian solutions have been done for this problem in the case of dependent processes.

A different approach for the change points' finding appears in machine learning literature, where the problem is formulated as unsupervised clustering of a set of temporal observations, for the identification of homogeneous sequences with respect to a given distance measure (Minkowski or Dynamic Time Warping). Common methods for finding the right segmentation consider top–down, bottom–up or sliding window algorithms [7, 83]. The reader may refer for an overview of these techniques to [76, 120].

Last period, the breakpoints' detection is done in hydrological applications by using CUSUM or segmentation procedures, as alternatives to the classical tests of Buishand, Lee and Heghinian, Pettitt.

CUSUMs are relatively simple tests based on the cumulative sums charts, which have the advantage to offer graphical interpretations of the results. Parametric CUSUMs are based on the comparison of the probability distribution functions before and after the changing moment, in the hypothesis of data independence. The nonparametric CUSUM test is a rank-based method that allows comparisons of successive observations with the median of the series for detecting a change in its mean after some observations. The test statistic is the cumulative sum of the k signs of the difference from the median. It is more robust to the autocorrelation existence than the parametric one [47]. A version of it is implemented in Change Point Analyzer software.

Different authors used segmentation procedures. Liu [84] performed the breakpoint analyzes for the segmentation of the seasonal runoff in many seasons, and his technique was validated by the Monte–Carlo method. Duggins [31] proposed an alternative method for the breakpoints detection, using transition matrices. Tsakalias and Koutsoyiannis [118] developed a heuristic algorithm, which emulates the exploratory data analysis of the human expert and which encodes a number of search strategies in a pattern directed computer program. Hubert segmentation procedure [68, 69] was generalized by Kehagias et al. [75]. They proposed a dynamical programming solution to the problem: divide a given time series into homogeneous segments so that the contiguous segments are heterogeneous. A new segmentation algorithm for long hydro-meteorological time series that combines the dynamic programming with the remaining cost concept has been proposed by Gedikli et al. [40, 41]. Also, a user-friendly program (*segmenter*) has been built to perform segmentation-by-constant and segmentation-by-linear-regression.

Whatever the procedure for the change point detection is, the null hypothesis (H_0) is that the series has no breakpoint, and its alternative (H_1) is that the series has at least a breakpoint.

Change points tests are implemented in different software. We discuss here only the capabilities of Khronostat and R.

Using Khronostat one can perform the change point detection by four different methods: Buishand [17, 18], Lee and Heghinian [79], Pettitt [99] and Hubert [68, 69]. A short description of these tests is done in [7]. We mention that the Buishand and Lee & Heghinian tests are based on the hypothesis of series' normality. The nonparametric test of Pettitt can be employed even if the series distribution is unknown, in the independence hypothesis. Among them, only the segmentation procedure of Hubert detects multiple breaks and the moments of their apparition, based on the Scheffé test [108].

The results of these tests for Constanta annual series (1961–2013), at 0.95 confidence level, are: Buishand doesn't reject the null hypothesis (Fig. 4), Pettit, Lee & Heghinian (Fig. 5) and Hubert segmentation procedure reject it. The last three tests indicate 1994 as a change point.

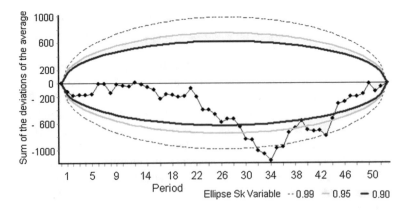

Fig. 4 Bois' ellipse associated with the Buishand test

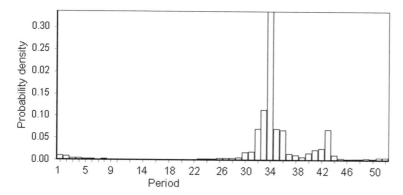

Fig. 5 A posteriori probability density of a break time position

The mDP algorithm, implemented in *segmenter*, based on the Scheffé test for the breakpoints selection, provides the same breakpoint—1994 (corresponding to 34 on the abscissa) (Fig. 6).

There are many packages in R software that perform the breakpoints detection, as: **bcp, ecp, changepoint, cpm, BFAST, SLC, strucchange, wbs** etc. In the following we present the capabilities of the first four packages.

The **bcp** package [37] provides an implementation of the Bayesian change point procedure of Barry and Hartigan [9] for univariate time series. It employs the Markov Chain Monte Carlo method and provides for each time moment the posterior probability of a change and the posterior mean. The procedure assumes that the observations are independent, normally distributed, with the same variance, but the independence hypothesis could be weakened.

We present here the code for the change point detection for annual precipitation series registered at Constanta, using the **bcp** package.

```
library(bcp)
data<- read.csv("D:\\Lucrari_2.12.14\\2015_Carte\\Cta_annual_1961_2013.csv", sep=",",
header=TRUE)
X<-data[,1]
Z<-as.vector(X)
bcp.0 <- bcp(Z)
plot.bcp(bcp.0) # produces Posterior Means: location in the sequence versus the posterior
# means over the iterations; Posterior Probability of a Change: location in the sequence
# versus the relative frequency of iterations which resulted in a change point – Fig. 7
bcp(Z, w0 = 0.2, p0 = 0.2, burnin = 50, mcmc = 500, return.mcmc = FALSE)
# returns the Bayesian change point summary (probability of a change in mean and posterior
# means
fitted.bcp(bcp.0)  # returns fitted values extracted from the bcp object
residuals.bcp(bcp.0) # returns residual extracted from the bcp object
```

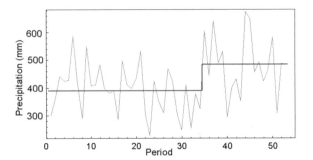

Fig. 6 Breakpoint selection by mDP algorithm with constant regression, implemented in *segmenter* software

In Fig. 7 the upper part presents the posterior means calculated for each point in the data series, and the lower part, the posterior probability that the point is a change one. The posterior probability of changes is not significant, excepting for the years 1994 and 2003 (corresponding to 34 and 43 on the abscissa).

The **ecp** package [57] is designed to determine (multiple) distributional change points and their locations in univariate and multivariate time series, while making the assumptions that the observations are independent over time, and there is $\alpha \in (0, 2]$ such that the absolute moment of order α of the series' distribution exists.

Divisive or agglomerative algorithms are the base of hierarchical estimation.

Divisive estimation sequentially identifies change points using a bisection algorithm. The agglomerative algorithm estimates change point locations looking for optimal segmentation. Both approaches detect changes in the data distribution. In the following we present the code for the divisive algorithm, applied to Constanta monthly series (1961–2013).

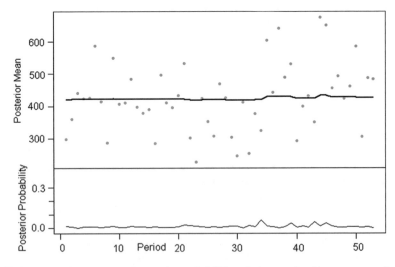

Fig. 7 Bcp: posterior means and posterior probabilities of changes for Constanta annual series (1961–2013)

```
library(ecp)
data<- read.csv("D:\\Lucrari_2.12.14\\2015_Carte\\Cta_lunar_1961_2013.csv", sep=",",
header=TRUE)
W<-data[,1]
Z1<-matrix(W, ncol=1)
Z1
output1 <- e.divisive(Z1, R = 499, alpha = 1)
output1
```

```
$k.hat
[1] 1

$order.found
[1]   1 637

$estimates
[1]   1 637

$considered.last
[1] 406

$p.values
[1] 0.146

$permutations
[1] 499

$cluster
  [1] 1 1 1 1 1 1 1 1 1 1 1 1 1 1 1 1 1 1 1 1 1 1 1 1 1 1 1 1 1 1 1 1 1 1 1 1 1 1
 [39] 1 1 1 1 1 1 1 1 1 1 1 1 1 1 1 1 1 1 1 1 1 1 1 1 1 1 1 1 1 1 1 1 1 1 1 1 1 1
 [77] 1 1 1 1 1 1 1 1 1 1 1 1 1 1 1 1 1 1 1 1 1 1 1 1 1 1 1 1 1 1 1 1 1 1 1 1 1 1
[115] 1 1 1 1 1 1 1 1 1 1 1 1 1 1 1 1 1 1 1 1 1 1 1 1 1 1 1 1 1 1 1 1 1 1 1 1 1 1
[153] 1 1 1 1 1 1 1 1 1 1 1 1 1 1 1 1 1 1 1 1 1 1 1 1 1 1 1 1 1 1 1 1 1 1 1 1 1 1
[191] 1 1 1 1 1 1 1 1 1 1 1 1 1 1 1 1 1 1 1 1 1 1 1 1 1 1 1 1 1 1 1 1 1 1 1 1 1 1
[229] 1 1 1 1 1 1 1 1 1 1 1 1 1 1 1 1 1 1 1 1 1 1 1 1 1 1 1 1 1 1 1 1 1 1 1 1 1 1
[267] 1 1 1 1 1 1 1 1 1 1 1 1 1 1 1 1 1 1 1 1 1 1 1 1 1 1 1 1 1 1 1 1 1 1 1 1 1 1
[305] 1 1 1 1 1 1 1 1 1 1 1 1 1 1 1 1 1 1 1 1 1 1 1 1 1 1 1 1 1 1 1 1 1 1 1 1 1 1
[343] 1 1 1 1 1 1 1 1 1 1 1 1 1 1 1 1 1 1 1 1 1 1 1 1 1 1 1 1 1 1 1 1 1 1 1 1 1 1
[381] 1 1 1 1 1 1 1 1 1 1 1 1 1 1 1 1 1 1 1 1 1 1 1 1 1 1 1 1 1 1 1 1 1 1 1 1 1 1
[419] 1 1 1 1 1 1 1 1 1 1 1 1 1 1 1 1 1 1 1 1 1 1 1 1 1 1 1 1 1 1 1 1 1 1 1 1 1 1
[457] 1 1 1 1 1 1 1 1 1 1 1 1 1 1 1 1 1 1 1 1 1 1 1 1 1 1 1 1 1 1 1 1 1 1 1 1 1 1
[495] 1 1 1 1 1 1 1 1 1 1 1 1 1 1 1 1 1 1 1 1 1 1 1 1 1 1 1 1 1 1 1 1 1 1 1 1 1 1
[533] 1 1 1 1 1 1 1 1 1 1 1 1 1 1 1 1 1 1 1 1 1 1 1 1 1 1 1 1 1 1 1 1 1 1 1 1 1 1
[571] 1 1 1 1 1 1 1 1 1 1 1 1 1 1 1 1 1 1 1 1 1 1 1 1 1 1 1 1 1 1 1 1 1 1 1 1 1 1
[609] 1 1 1 1 1 1 1 1 1 1 1 1 1 1 1 1 1 1 1 1 1 1 1 1 1 1 1 1 1
```

changepoint is a complete R package, which provides the implementation of different methods for the detection of changes in mean and variance. Here we present only some procedures treating the change points in mean.

'binseg.mean.cusum' is an approximate method that uses the Binary Segmentation [109] to calculate the number of breakpoints and their position for the cumulative sums test statistic. The first argument of the function is a vector that contains the series values; the second one is the maximum number of breakpoints (fixed by the user) and the third one, the value of the penalty function. This algorithm can be utilized without restrictive assumptions on the dataset distribution.

For example, for Constanta annual series (1961–2013):

```
library(changepoint)
data<- read.csv("D:\\Lucrari_2.12.14\\2015_Carte\\Cta_annual_1961_2013.csv", sep=",",
header=TRUE)
x<-data[,1]
binseg.mean.cusum(x, Q=2, pen=0.8) # looks for two change change points


    $cps
     [,1]        [,2]
[1,] 34.00000    21.00000
[2,]  21.81962    18.64178

    $op.cpts # gives the optimal change point locations for the penalty supplied,
    [1] 2      # that is 34, in our case.

    $pen
    [1] 0.8
```

If data is normally distributed, 'binseg.mean.cusum()' may be replaced by the function 'binseg.mean.norm' with the same arguments.

Another possibility is the use of 'cpt.mean()', with the arguments:

- Data—contains the data within which we look for the breakpoint;
- Penalty—with the possibilities of choice of None, SIC, BIC, AIC, Hannan–Quinn, Asymptotic, Manual;
- pen.value—a numeric value of penalty, when choosing "Manual", or a text that gives the formula to use, as: n—sample volume, null—null likelihood, alt—alternative likelihood, tau—proposed change point, diffparam–difference in number of alternative and null parameters;
- method—'AMOC' (for single changepoint), 'BinSeg' [109], 'SegNeigh' [5] or 'PELT' [77] (for multiple change points);
- Q—maximum number of change points when 'BinSeg' is selected or maximum number of segments when 'SegNeigh' is used;
- test.stat—the test statistic or the data distribution: 'Normal' or 'CUSUM';
- class—if TRUE, then an object of class 'cpt' is returned;
- param.estimates—if TRUE and class is also TRUE, then parameter estimates are returned. Otherwise, no parameter estimate is returned.

Some examples are presented in the following:

```
library(changepoint)
data<- read.csv("D:\\Lucrari_2.12.14\\2015_Carte\\Cta_annual_1961_2013.csv", sep=",",
header=TRUE)
x<-data[,1]
cpt.mean(x, penalty="SIC", method="AMOC", class=FALSE)

       cpt p    value
       34       1
# only a change point was detected in the data series and this is the 34-th value

ans=cpt.mean(x, penalty="Asymptotic", pen.value=0.01, method="AMOC")

       "cpttype":
       [1] "mean"

       Slot "method":
       [1] "AMOC"

       Slot "test.stat":
       [1] "Normal"

       Slot "pen.type":
       [1] "Asymptotic"

       Slot "pen.value":
       [1] 27.97932

       Slot "cpts":
       cpt
       34  53

       Slot "ncpts.max":
       [1] 1

       Slot "param.est":
       $mean
       [1] 389.5324 484.4105
```

ans = cpt.mean(x, penalty="Manual", pen.value=0.8, method="AMOC", test.stat="CUSUM")
ans

An object of class "cpt"
Slot "data.set":
Time Series:
Start = 1
End = 53
Frequency = 1
 [1] 299.2 361.0 440.9 424.2 426.1 586.2 415.1 288.4 548.6 407.4 410.9 483.8
[13] 397.3 379.1 390.1 285.0 496.6 411.4 396.4 433.9 532.4 301.9 227.0 424.9
[25] 352.3 308.8 469.3 425.6 305.3 246.6 412.3 253.8 378.2 324.1 604.3 443.3
[37] 641.2 488.8 531.1 292.5 400.4 430.6 350.2 674.6 649.9 456.2 493.7 423.9
[49] 461.7 583.8 307.0 487.6 483.0

Slot "cpttype":
[1] "mean"

Slot "method":
[1] "AMOC"

Slot "test.stat":
[1] "CUSUM"

Slot "pen.type":
[1] "Manual"

Slot "pen.value":
[1] 0.8

Slot "cpts":
cpt
35 53 # change points detected: the 35th and the 53rd values in the series

Slot "ncpts.max":
[1] 1

Slot "param.est":
$mean
[1] 395.6686   477.7500 # the values of the series in the change points.

ans = cpt.mean(x, penalty="AIC", method="SegNeigh", test.stat="Normal")
print(ans) # print details of the methods and a summary of results

summary(.) :
Changepoint type   : Change in mean
Method of analysis   : SegNeigh
Test Statistic   : Normal
Type of penalty   : AIC with value 2
Maximum no. of cpts : 5
Changepoints location : 34, 39, 43, 45

plot(ans) # plots the data and the change point, as in **Fig. 8.**

Fig. 8 Output of 'plot(ans)' command

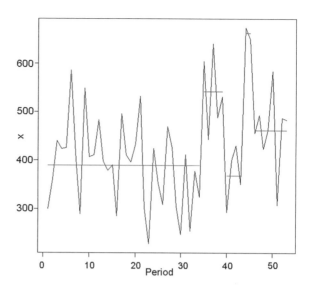

Other options offered by this package are: 'single.mean.cusum.calc' 'multiple. mean.cusum', 'multiple.mean.norm' etc. The first one calculates the CUSUM test statistic for all possible breakpoints' locations and provides only the most probable one. Even if it can be used if no assumption on the data distribution is done, there is no test implemented to confirm that the change point detected is a true breakpoint, so the test must be utilized with cautions. For example:

```
library(changepoint)
data<- read.csv("D:\\Lucrari_2.12.14\\2015_Carte\\Cta_annual_1961_2013.csv", sep=",",
header=TRUE)
  x<-data[,1]
  single.mean.cusum.calc(x,extrainf=TRUE)

        cpt test        statistic
        35.00000        21.81962
```

'multiple.mean.cusum.calc' has the arguments:

- data—contains the data within which we look for the breakpoint;
- mul.method—'BinSeg' or 'SegNeigh'
- penalty—with the possibilities of choice of: None, SIC, BIC, AIC, Hannan–Quinn, Asymptotic, Manual;
- pen.value—a numeric value of penalty, when choosing "Manual", or a text that gives the formula to use, as: n—sample volume, null—null likelihood, alt—

alternative likelihood, tau—proposed change point, diffparam–difference in number of alternative and null parameters;

- Q—maximum number of change points when 'BinSeg' is selected or maximum number of segments when 'SegNeigh' is used;
- class—if TRUE, then an object of class 'cpt' is returned;
- param.estimates—if TRUE and class is also TRUE, then parameter estimates are returned. Otherwise, no parameter estimate is returned.

For example:

```
library(changepoint)
data<- read.csv("D:\\Lucrari_2.12.14\\2015_Carte\\Cta_annual_1961_2013.csv", sep=",",
header=TRUE)
x<-data[,1]
multiple.mean.cusum(x, mul.method="SegNeigh", penalty="AIC", pen.value= null, Q=5,
class=FALSE, param.estimates=TRUE)
```

$cps

	[,1]	[,2]	[,3]	[,4]	[,5]
[,1]	0	0	0	0	0
[,2]	34	0	0	0	0
[,3]	34	21	0	0	0
[,4]	43	39	34	0	0
[,5]	45	43	39	34	0

$op.cpts # gives the optimal change point locations for the penalty supplied
[1] 4

$pen # provides the penalty used for finding the optimal change points number
[1] 2

Another complex package is **cpm** [53]. This framework is an approach to Phase II process monitoring (also known as sequential change detection) for performing breakpoints detection, using parametric and nonparametric procedures, on univariate data streams.

In Batch detection (also known as Phase I) the researcher has to look for the change points, using all the available observations in a data set with a fixed length. In Sequential detection, the observation is processed at the moment of its apparition and the decision concerning the break occurrence is based only on the previously received data. When a change point is detected, the change detector is restarted from the following observation in the sequence, allowing the detection of multiple breakpoints.

The procedures implemented in cpm permit single or multiple change points detection in mean or variance, in Gaussian or non-Gaussian series, using the statistics: Student, Bartlett, GLR, Exponential, GLR Adjusted, Exponential Adjusted, FET,

Mann–Whitney, Mood, Lepage, Kolmogorov–Smirnov and Cramer-von-Mises. The first three statistics are used for detection of changes in a Gaussian sequence, respectively in mean, in variance, in mean and variance. The fourth and fifth test statistics are used to determine changes of the parameter of an exponentially distributed data series; the sixth one performs the Fisher test for changes in the parameter of a Bernoulli distributed sequence. The last five tests can be used if the process generating the data is not Gaussian. They are respectively [107]:

- the Mann–Whitney test, for detecting the position of the shifts in a stream;
- the Mood test, for detecting the shifts in a stream;
- the Lepage test, for detecting the position and/or the shifts in a stream;
- the Kolmogorov–Smirnov and Cramer-von-Mises tests, for detection of arbitrary changes in a stream.

In Phase I, 'detectChangePointBatch' function is used for testing the hypothesis of the existence of a single change point in a series of observations, and estimating its location. For example, for Constanta annual precipitation series the application of Kolmogorov–Smirnov method is done by writing the following commands:

```
data<- read.csv("D:\\Lucrari_2.12.14\\2015_Carte\\Cta_annual_1961_2013.csv", sep=",",
header=TRUE)
    x <- data[,1]
    resultsKS <- detectChangePointBatch(x, cpmType = "Kolmogorov-Smirnov", alpha =
0.05) # detects the change points by the Kolmogorov-Smirnov method
    plot(x, type='l')
    if (resultsKS$changeDetected) {
    abline(v = resultsKS$changePoint, lty=2)
    }
# plots the data series and marks the moment of the break with a vertical line (Fig. 9)
    resultsKS


        $changePoint
        [1] 34

        $changeDetected
        [1] TRUE

        $alpha
        [1] 0.05

        $threshold
        [1] 0.9966167
```

Fig. 9 Breakpoint detected in Constanta annual series (1961–2013) by the Kolmogorov–Smirnov method (Phase I)

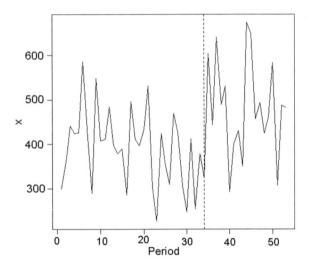

'detectChangePoint' allows performing Phase II analysis. For example, for Constanta monthly series (1961–2013), we have the following commands:

```
data<-read.csv("D:\\Lucrari_2.12.14\\2015_Carte\\Cta_lunar_1961_2013.csv",    sep=",",
header=TRUE)
 x <- data[,1]
 resultsLepage <- detectChangePoint(x, cpmType = "Lepage", ARL0=500)
 # detects the change points by the Mood method;
 # ALRO ∈ {370, 500, 600, 700, … , 1000, 2000, … , 10000, 20000, 50000}
 plot(x, type='l')
 if (resultsLepage$changeDetected) { abline(v = resultsLepage$detectionTime, lty=2)}
 resultsLepage

     $changePoint # change point estimated
     [1] 435

     $detectionTime # the moment at which the change was detected
     [1] 440

     $changeDetected
     [1] TRUE
```

In the case when the stream of observations contains many change points, the 'processStream' function can be used for their detection because it allows processing the observations one-by-one. For example:

Fig. 10 Changes in mean for Constanta monthly series. The two *solid black lines* indicate the moments when the changes were detected, and the *dotted lines* indicate the estimated change points' locations

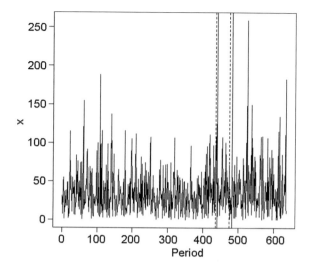

```
    data<-read.csv("D:\\Lucrari_2.12.14\\2015_Carte\\Cta_lunar_1961_2013.csv",    sep=",",
header=TRUE)
    x <- data[,1]
    res <- processStream(x, cpmType="Mann-Whitney", ARL0=500, startup=20)
# startup represents the number of observations after which the monitoring  begins
    res
```

```
        $changePoints #the estimated change points
        [1] 435 474

        $detectionTimes
        [1] 440 481
```

```
    plot(x,type='l')
    abline(v=res$detectionTimes) #draws vertical lines with the abscissa equal to the
# moments when the change points were detected
    abline(v=res$changePoints, lty=2) #draws vertical lines with the abscissa equal to the
# estimated moments of changes – Fig. 10.
```

6   Testing for Long Range Dependence Property

By definition, a time series has the long range dependence (LRD) property if the series $\sum_{h=-\infty}^{\infty} \rho(h)$ is divergent.

This definition is equivalent to the following behavior of the series' autocovariance: $\gamma(k) \sim k^{-\alpha}L(k)$, where $0 < \alpha < 1$ and $L(k)$ is a function with a slow variation at infinity.

LRD manifests in the time domain as a high level of correlation between points situated at big distances in time and, in the frequency domain, as a significant level of power at frequencies near zero [36].

A measure for long term memory (and fractality) of a time series is the Hurst exponent, H [88]. Its values range in the interval $(0, 1)$. The value of H in the interval $(0, 0.5)$ indicates an anti-persistent behavior of a time series, which is an up value is more likely followed by a down value, and vice versa. $H = 0.5$ indicates a random series. $H \in (0.5, 1)$ corresponds to a persistent series, that is the next value has more likely the same direction as the current one.

Different methods have been developed to carry out the LRD analysis: R/S [70] and Lo's modified R/S method [86], Aggregated Variance Method [117], Absolute values of aggregated series [117], Ratio of Variance of Residuals [117], Periodogram [43], Higuchi Method [50], Detrended Fluctuation Analysis—DFA [97, 98], Whittle's approximate MLE and Local Whittle estimators [104], methods based on wavelets etc. The reader might also refer to [117].

We shortly describe some of them here. For this aim we denote by $(x_k)_{k \in \overline{1,n}}$ the studied data series.

In the rescaled analysis (R/S method), the data series is divided in d sub-series of the same length. For each sub-series, the mean and standard deviation are computed, the data are normalized by subtracting the sub-series average, the cumulative sub-series is created by adding up the normalized values, the range is determined as difference between its maximum and minimum, and the rescaled range is computed dividing the sub-series range to its standard deviation. Finally, the average of the rescaled ranges (R/S) of all subseries is determined.

The procedure is repeated for different d.

Hurst coefficient is determined as the slope of the fitted line of log(R/S) on log(d).

Different corrections have been proposed by Lo, Andrews, Wang, etc. but we shall not discuss them here.

The function 'hurstexp' from the package **pracma** [65] in R calculates the Hurst exponent using the R/S method and some corrections of it, based on the article [121]. Here is an example:

```
data<-read.csv("D:\\Lucrari_2.12.14\\2015_Carte\\Cta_lunar_1961_2013.csv", sep=",",
header=TRUE)
    x <- data[,1]
    library(pracma)
    hurstexp(x, d = 50, display = TRUE) # d is the smallest box size - default is 50;
```

Simple R/S Hurst estimation:	0.582716
Corrected R over S Hurst exponent:	0.6345952
Empirical Hurst exponent:	0.5824732
Corrected empirical Hurst exponent:	0.5368507
Theoretical Hurst exponent:	0.5444537

The rescale analysis (R/S) can also be performed using 'rsFit' function from **fArma** package [54] of R. For example:

```
data<-read.csv("D:\\Lucrari_2.12.14\\2015_Carte\\Cta_lunar_1961_2013.csv", sep=",",
header=TRUE)
  x <- data[,1]
  library(fArma)
  rsFit(x, levels = 50, minnpts = 3, cut.off = 10^c(0.7, 2.5), doplot = FALSE, trace =
FALSE, title = NULL, description = NULL)


    Title:
    Hurst Exponent from R/S Method

    Call:
      rsFit(x = x, levels = 50, minnpts = 3, cut.off = 10^c(0.7, 2.5),  doplot = FALSE,
    trace = FALSE, title = NULL, description = NULL)

    Method:
    R/S Method

    Hurst Exponent:
      H          beta
    0.5785334 0.5785334

    Hurst Exponent Diagnostic:
      Estimate      Std.Err     t-value      Pr(>|t|)
    X  0.5785334 0.02859643 20.23096  2.835059e-23

    Parameter Settings:
      n    levels  minnpts cut.off1 cut.off2
      636    50      3        5       316
```

In *Aggregated variance method*, after dividing the original data series into d subseries of the same length, the aggregated series is formed by the averages of the values in the sub-series, and the sample variance is computed. For different values of d, the sample variance is plotted against d on a log-log scale and a least squares line to the points of the plot is drawn, whose slope, β, is used for computing the Hurst coefficient. For fGN and ARIMA processes, $-1 < \beta = 2H - 2 < 0$.

In *Absolute values of the aggregated series method*, the aggregated series is formed as in Aggregated variance method, then the n-th absolute value of the resulted sub-series are computed instead of their variances. For different values of d, the n-th absolute value is plotted against d on a log-log scale and a least squares line to the points of the plot is drawn, whose slope, β, is used for computing the Hurst coefficient.

To apply the *Differenced variance method*, one has to pass through the stages:

- Produce the aggregated series of order m, for various m;
- Find the sample variance for each m;

- Plot the log(variance) as a function of $\log(m)$;
- If there are suspicions about the non-stationary existence, differentiate the variances, as a function of m;
- Plot the results on a log–log plot.

If there are shifts in the mean or a slowly declining trend is present, then the plot from the third step should be in an exponential form, and that from the last one, in a linear form, with the slope $\beta = 2H - 2$.

The use of this technique is recommended to distinguish the cases: (a) H is near 0.5 and there are jumps in the mean; (b) H is significantly larger than 0.5, and there is a non-zero trend [116].

Higuchi's method is based on dividing a given data series into sub-series, defining the length of the curve for an interval and detecting its fractal dimension. More precisely, the steps are:

- From the data series $(x_i)_{i=\overline{1,n}}$, built the sub-series (X_k^m) containing $(x_m, x_{m+k}, x_{m+2k}, \ldots, x_{m+[(n-m)/k] \cdot k})$, where [] denotes the integer part;
- Define the length of X_k^m as:

$$
L_m(k) = \frac{1}{k} \left\{ \sum_{i=1}^{\left[\frac{n-m}{k}\right]} \left| x_{m+ik} - x_{m+(i-1)k} \right| \cdot \frac{n-1}{\left[\frac{n-m}{k}\right] \cdot k} \right\},
$$

- Define the length of the curve for the time interval k, $\langle L(k) \rangle$, as the mean value over k sets of $L_m(k)$.

If $\langle L(k) \rangle \sim k^{-D}$, the curve is fractal with the dimension D.

'hurstBlock' from **fractal** package [56] of R is a function used for estimating the Hurst exponent of a long memory time series, by choosing one of the methods specified in its argument. All implemented procedures work on the brute series, not on its spectrum.

The arguments of "hurstBlock" are:

- x—a vector containing the values of a time series, uniformly-sampled.
- fit—a function giving the linear regression method used for fitting the statistics (on a log-log scale). It can be: 'lm', 'lmsreg', and 'ltsreg', the default being 'lm'.
- method—a character string that indicates the method used for estimating the Hurst coefficient. The default is: "aggabs" (Absolute Values of the Aggregated Series), but one can also choose:

  - "aggvar"—Aggregated Variance Method [117]
  - "diffvar"—Differenced Variance Method [117]
  - "higuchi"—Higuchi's Method [50]

- scale.max—the maximum block size used in partitioning the data series. The default is 'length(x)'.

- scale.min—the minimum block size used in partitioning the data series. The default is 8.
- scale.ratio—ratio of successive scales used in partitioning the data series. The default is 2.
- Weight—a function of a variable (x) used for weighting the resulting statistics (x) for each scale during the linear regression. It is supported when fit = lm. The default is 'function(x) rep(1,length(x))'.

For example, to calculate the Hurst coefficient for Constanta monthly series using Higuchi's method, the following code must be written:

```
data<-read.csv("D:\\Lucrari_2.12.14\\2015_Carte\\Cta_lunar_1961_2013.csv", sep=",",
header=TRUE)
x <- data[,1]
library(fractal)
y<-hurstBlock(x, method="higuchi", scale.min=8, scale.max=636, scale.ratio=2,
weight=function(x) rep(1,length(x)), fit=lm)
y
```

```
H estimate        : 0.9949413
Domain            : Time
Statistic         : higuchi
Length of series  : 636
Block overlap fraction: 0
Scale ratio       : 2

Scale      8.0    16.0   32.00   64.00  128.00  256.000  512.000
higuchi  2798.2  1401.1  702.65  351.19  173.58   84.716   43.431
```

```
# plots the results
plot(u, key=FALSE)
mtext(paste("higuchi", round(as.numeric(u),3), sep=", H="), line = 0.5, adj=1)
```

The same methods are implemented in **fArma** package from R.

The sequence of commands for performing them, in the case of Constanta monthly series, and the results are:

```
data<-read.csv("D:\\Lucrari_2.12.14\\2015_Carte\\Cta_lunar_1961_2013.csv", sep=",",
header=TRUE)
  x <- data[,1]
  aggvarFit(x, levels = 50, minnpts = 3, cut.off = 10^c(0.7, 2.5), doplot = TRUE, trace =
FALSE, title = NULL, description = NULL)
```

Method:
Aggregated Variance Method
Hurst Exponent:
         H   beta
   0.6088583   -0.7822834

Hurst Exponent Diagnostic:
      Estimate   Std.Err   t-value   Pr(>|t|)
  X 0.6088583   0.03364099   18.0987   4.414577e-23

```
  absvalFit(x, levels = 50, minnpts = 3, cut.off = 10^c(0.7, 2.5), moment = 1, doplot =
TRUE, trace = FALSE, title = NULL, description = NULL)
```

Method:
Absolute Moment - No. 1
Hurst Exponent:
         H   beta
   0.6749373   -0.3250627

Hurst Exponent Diagnostic:
      Estimate   Std.Err   t -value   Pr(>|t|)
  X 0.6749373   0.03679841   18.34148   2.520645e-23

```
  diffvarFit(x, levels = 50, minnpts = 3, cut.off = 10^c(0.7, 2.5), doplot = TRUE, trace =
FALSE, title = NULL, description = NULL)
```

Method:
Differenced Aggregated Variance
Hurst Exponent:
         H   beta
   0.7531249   -0.4937501

Hurst Exponent Diagnostic:
      Estimate   Std.Err   t-value   Pr(>|t|)
  X 0.7531249   0.1412358   5.332396   2.052936e-05

Parameter Settings:
    n   levels minnpts cut.off1 cut.off2
   636   50   3   5   316

```
  higuchiFit(x, levels = 50, minnpts = 2, cut.off = 10^c(0.7, 2.5), doplot = TRUE, trace =
FALSE, title = NULL, description = NULL)
```

Method:
Higuchi Method
Hurst Exponent:
         H   beta
   0.9578126   -1.0421874

Hurst Exponent Diagnostic:
      Estimate   Std.Err   t-value   Pr(>|t|)
  X 0.9578126   0.02840518   33.71965   4.447243e-35

The arguments of the previous functions are:

- x—the series to be analyzed;
- level—the number of blocks from which the statistics values are computed;
- minnpts—the minimum number of points or block size used;
- cut.off—a vector containing the lower and upper cut of points. The default is c (0.7, 2.5);
- doplot—a logical value; if TRUE, a plot is displayed;
- trace—a logical value, by default FALSE. It indicates if the process is traced;
- title—a character string allowing for a title of the project;
- description—a character string allowing for a description.

The results depend on the cut.off and are plotted in Fig. 11.

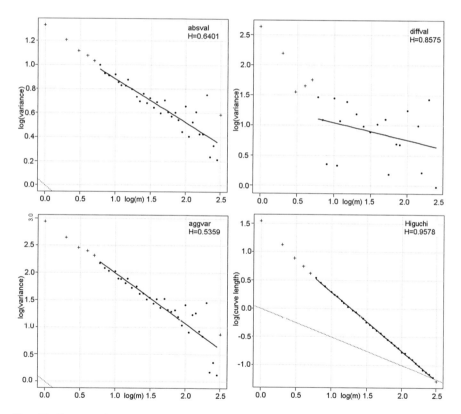

Fig. 11 Hurst coefficient computed by Absolute values, Aggregated variance, Differenced variance and Higuchi methods for Constanta monthly series

To perform *Detrended Fluctuation Analysis* [97, 98], the data series is firstly integrated and divided into sub-series of the same length, m, for which polynomial trends are fitted. The integrated series is detrended by subtracting from each sub-series the corresponding polynomial trend, and then the root mean-square ($F(m)$) of the new series is computed. These steps are repeated for different lengths of the sub-series, for providing a linear relationship between $F(m)$ and m.

'DFA' function from fractal package is employed for performing the Detrended Fluctuation Analysis. It is useful for analyzing the long-memory in data series whose spectral density function has the form $S(f) \sim f^{\alpha}$ at low frequencies, $f \in (0, 0.5)$ being the normalized frequency variable and $\alpha < -1$, the scaling exponent. If $\alpha > -1$, then cumulative summations of the data series must be performed for increasing the scaling exponent (each cumulative summation decreases the exponent by 2). The user may also use the differencing operation prior to the DFA analysis.

'DFA' function has the following arguments:

- x—a vector containing the values of a time series, uniformly-sampled;
- detrend—a character string that denotes the detrending type used on each sub-series. It can be a polynomial (for example, for the polynomial of first order the type is "poly1"), "bridge" or "none";
- overlap—the overlap of blocks in partitioning the time data expressed as a fraction in [0,1); the default is 0;
- scale.max—the maximum block size used in partitioning the series; the default is trunc(length(x)/2);
- scale.min—the minimum block size used in partitioning the data; the default is 2 (k + 1) for polynomial detrending (k is the degree of the polynomial) and it is min{4, length(x)/4} for the other detrending techniques;
- scale.ratio—the ratio of successive scales; the default is 2;
- sum.order—the number of differences or cumulative summations performed on the brute series prior to apply DFA. The default is 0, sum.order > 0 for cumulative summations, sum.order = p < 0 if a difference of p order is performed;
- verbose—a logical value, indicating if or not the detrending model and processing progress information is displayed (when it is TRUE/FALSE); the default is: FALSE.

For example, for Constanta monthly series:

```
data<-read.csv("D:\\Lucrari_2.12.14\\2015_Carte\\Cta_lunar_1961_2013.csv",sep=",",
header=TRUE)
x <- data[,1]
library(fractal)
DFA.x <- DFA(x, detrend="poly0", sum.order=0)
eda.plot(DFA.x) #plots the charts with the results
print(DFA.x) #prints the results - Fig. 12, the left hand side
```

```
Detrended fluctuation analysis for x
------------------------------------

H estimate              : 0.06846433
Domain                  : Time
Statistic               : RMSE
Length of series        : 636
Block detrending model : x ~ 1
Block overlap fraction   : 0
Scale ratio             : 2

Scale    2.000   4.000   8.000 16.000 32.000 64.000 128.000 256.000
RMSE  20.799 24.992 27.149 28.263 28.205 28.529  26.682   26.832
```

```
DFA.x <- DFA(x, detrend="poly0", sum.order=1)
print(DFA.x)
```

```
Detrended fluctuation analysis for x
------------------------------------

H estimate              : 0.968804
Domain                  : Time
Statistic               : RMSE
Length of series        : 636
Block detrending model: x ~ 1
Block overlap fraction   : 0
Scale ratio             : 2
Preprocessing           : 1st order cumulative summation

Scale    2.000 4.000 8.000  16.00  32.00  64.00 128.0  256.0
RMSE  22.879 45.425 87.928 169.43 339.58 652.68 1293.6 2538.7
```

```
eda.plot(DFA.x) #prints the results - Fig. 12, the right hand side
```

The function 'perFit' from fArma package may be employed for computing the Hurst exponent using the *periodogram method* [43], that is based on the estimation of spectral density of a time series (which is the periodogram, in the case of finite variance).

A series with long range dependence property has a spectral density with a power law behavior in frequency in the neighborhood of zero; therefore, for computational purposes only the lowest 10 % of the frequencies are used.

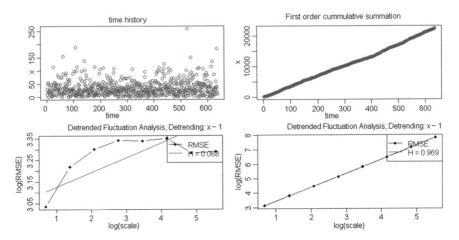

Fig. 12 The plot of DFA: the *left-hand side*, for the brute series, and the *right-hand side*, for the first order cumulative summation, applied to the brute series. The slope of the line which best fits a plot of log(RMSE) versus log(scale) is the scaling exponent

More precisely, the estimation of spectral density is determined by the least squares regression:

$$\ln(I_x(\omega_j)) = \alpha - d \ln[4 \sin^2(\omega_j/2)] + e_j,$$

where $I_x(\omega_j)$ is the sample periodogram at the j-th Fourier frequency $\omega_j = 2\pi j/T$ $(j = 1, \ldots, [T/2])$, e_j is the residual, that must be independent identically distributed, with a variance of $\pi^2/6$.

Therefore d in the previous equation is estimated by:

$$\hat{d}_{GPH} = \frac{\sum_{j=1}^{m} (Y_j - \bar{Y}) \ln\{I(\omega_j)\}}{\sum_{j=1}^{m} (Y_j - \bar{Y})^2}, \quad 0 < m < n,$$

where

$$Y_j = -\ln[4 \sin^2(\lambda_j/2)] \quad \text{and} \quad \bar{Y} = \frac{1}{m}\sum_{j=1}^{m} Y_j. \ [43]$$

Plotting the periodogram versus frequency, in log-log scale and fitting a straight line by the least squares method, its slope will be $1 - 2H$.

In the implementation of 'perFit', selecting the argument method = "per", one can vary the cut off and plot H versus the cut off to determine where the curve flattened, in order to estimate H. Alternatively selecting method = "cumper", the cumulative periodogram is used and the slope of the log-log fit is $2 - 2H$.

For example:

```
data<-read.csv("D:\\Lucrari_2.12.14\\2015_Carte\\Cta_lunar_1961_2013.csv",sep=",",
header=TRUE)
  x <- data[,1]
  library(fArma)
  perFit(x, cut.off = 0.1, method = "per",  doplot = FALSE, title = NULL, description =
NULL)


        Title:
        Hurst Exponent from Periodogram Method

        Call:
        perFit(x = x, cut.off = 0.1, method = "per", doplot = TRUE, title = NULL,
    description = NULL)

        Method:
        Periodogram Method

        Hurst Exponent:
        H           beta
        0.5654380   -0.1308759

        Hurst Exponent Diagnostic:
        Estimate    Std.Err       t-value      Pr(>|t|)
        X 0.565438   0.08496654   6.654831   9.675144e-09

        Parameter Settings:
        n     cut.off
        636   10
```

Karagiannis et al. [74] implemented some of the presented methods in a user-friendly software, called *Selfis*. As a confirmatory analysis, after the LRD study, they propose the use of bucket shuffling, a procedure introduced in [38]. It decouples the short-term and the long-range correlations by shuffling sub-series of a given series, followed by the investigation of autocorrelation function [73], in five stages:

(i)   The data series (x_i) is divided into k buckets of the same length, b;
(ii) A home is attached to each value, x_i. It is defined by the bucket with the number $H(i) = [i/b]$, where [] denotes the integer part;
(iii) The in-buckets are built by the couples (x_i, x_j) such that $H(i) = H(j)$;
(iv) The out-buckets pairs are built by the couples of values for which $H(i) \neq H(j)$. For such pairs, the corresponding offset is $|H(i) - H(j)|$;
(v)   A randomization is performed. It can be of three types: external, internal or two-levels. In external randomization the buckets' order is randomized, but their content is preserved. In internal randomization the buckets' order is preserved, but their content is randomized. In two-levels randomization, the

Table 2 Results of LRD analysis running Selfis software, for Constanta daily series before and after internal shuffling

Method	Initial series		Shuffled series	
	H	Correl coef (%)/ C.I (95 %)	H	Correl coef (%)/ C.I (95 %)
Aggregated variance	0.527	97.37	0.536	96.22
Absolute moments	0.472	94.13	0.480	93.46
R/S	0.585	99.88	0.568	99.87
Variance of residuals	0.721	96.49	0.672	98.67
Whittle	0.585	[0.575; 0.595]	0.505	[0.496; 0.515]

buckets are divided in atoms of the same size, followed by an external randomization of the blocks of atoms of each bucket.

(vi) The autocorrelation function of the randomized series is analyzed for deciding about the existence of LRD property, based on the fact that the autocorrelation function of the internally-randomized series preserves the same characteristics as those of the initial data series [38, 75].

Therefore, if the study series has LRD, the ACF of the internally-randomized series should show the same slowly-decreasing behavior as the original one.

For Constanta daily series record from the period 1961–2009, Selfis has been used for the estimation of Hurst's coefficients (H) by different methods, before and after internal shuffling.

The results are presented in Table 2, where 'Correl coef (%)' represents the correlation coefficients given as percentages and 'C.I (95 %)' represents the corresponding confidence interval, at the confidence level of 95 %. We remark that there is no significant difference between the Hurst coefficients before and after shuffling, so the series has not LRD.

The autocorrelograms' analysis (Fig. 13) confirms this assertion.

7 Goodness of Fit Tests

After data series modeling, one wants to check if the two data samples (registered and estimated by the model) come from the same distribution. Note that, in practice, the common distribution is, usually, unknown. Therefore, one wants to test the null hypothesis:

(H$_0$) The two samples come from a common distribution, e.g. $\{F(z) = G(z), \forall z\}$, against the two-sided alternative:

(H$_1$): The two samples do not come from a common distribution, e.g.

$$\{\exists z: F(z) \neq G(z)\},$$

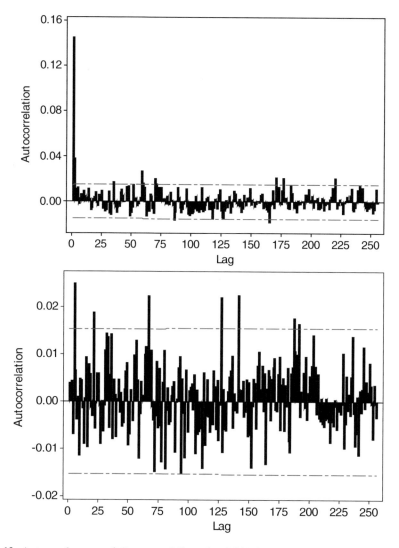

Fig. 13 Autocorrelograms of Constanta daily series (1961–2009) before and after the internal shuffling

or against one of the one-sided alternatives:

$$(H_1): \{\exists z: F(z) < G(z)\}, \{\exists z : F(z) > G(z)\},$$

where F and G are the continuous distribution functions.

The non-parametric test used for this purpose is called *the Kolmogorov-Smirnov test for two samples* and it does not rely on the normality assumption [25].

The implementation of this test is based on the computation of empirical cumulative distribution function (ecdf), and is done in R by using the function ecdf (x). To plot it, one can use the R command plot.ecdf(x).

It is possible to compare two independent samples, *x*, *y* by plotting their ecdf on the same chart, using the same scale. For example, if one wants to compare the samples formed respectively by the first 20 annual precipitation values registered at Constanta and the next 32 annual precipitation values, he uses the commands:

```
data<-read.csv("D:\\Lucrari_2.12.14\\2015_Carte\\Cta_annual_1961_2013.csv", sep=",",
header=TRUE)
    x<-data[1:20,1] # builds the first sample, containing the first 20 values of annual series
    x


    [1]   299.2 361.0 440.9 424.2 426.1 586.2 415.1 288.4 548.6 407.4 410.9 483.8
    [13] 397.3 379.1 390.1 285.0 496.6 411.4 396.4 433.9


    y<-data[21:53,1] # builds the second sample, containing the last 33 values of annual
# series
    y


    [1]   532.4 301.9 227.0 424.9 352.3 308.8 469.3 425.6 305.3 246.6 412.3 253.8
    [13] 378.2 324.1 604.3 443.3 641.2 488.8 531.1 292.5 400.4 430.6 350.2 674.6
    [25] 649.9 456.2 493.7 423.9 461.7 583.8 307.0 487.6 483.0


plot.ecdf(x,y, pch= "*")  # plots ecdf(x) using the scale of ecdf(y) - Fig. 14.


ks.test(x,y, alternative=c("two.sided", exact=NULL)) # performs the two-sample Kolmogorov-
# Smirnov test. NULL means that the exact p-value is computed.


        Two-sample Kolmogorov-Smirnov test
        data:  x and y
        D = 0.2545, p-value = 0.3301
        alternative hypothesis: two-sided.
```

So, the null hypothesis cannot be rejected.

If one of the distributions is known, the Anderson–Darling and Cramer-von Mises tests could be successfully used.

Fig. 14 Plot of ecdf(x) using the scale of ecdf(y)

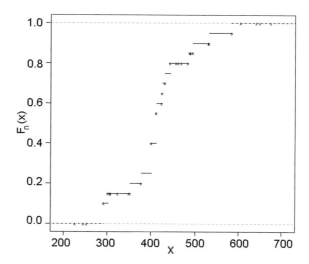

References

1. Aggarwal, C.C.: Outlier Analysis. Springer, New York (2013)
2. Anderson, T.W., Darling, D.A.: Asymptotic theory of certain "goodness of fit" criteria based on stochastic processes. Ann. Math. Stat. **23**, 193–212 (1952)
3. Arnold, B.T., Emerson, J.W.: Nonparametric goodness-of-fit tests for discrete null distributions. R J. **3**(2), 34–49 (2011)
4. Arranz, M.: Portmanteau test statistics in time series. http://www.tol-project.org/
5. Auger, I.E., Lawrence, C.E.: Algorithms for the optimal identification of segment neighborhoods. Bull. Math. Biol. **51**(1), 39–54 (1989)
6. Bakirov, N.K., Rizzo, M.L., Székely, G.J.: A multivariate nonparametric test for independence. J. Multivar. Anal. **97**(8), 1742–1756 (2006)
7. Barbulescu, A., Maftei, C., Bautu, E.: Modeling the Hydro-Meteorological Time Series. Applications to Dobrudja Region. LAP LAMBERT Academic Publishing GmbH & Co., Saarbrucken (2010)
8. Barnett, V., Lewis, T.: Outliers in Statistical Data. Wiley, New York (1994)
9. Barry, D., Hartigan, J.A.: A Bayesian analysis for change point problems. J. Am. Stat. Assoc. **88**, 309–319 (1993)
10. Bartlett, M.S.: Properties of sufficiency and statistical tests. Proc. Royal Stat. Soc. Series A **160**, 268–282 (1937)
11. Basseville, M., Nikiforov, I.: Detection of abrupt changes: theory and application. Prentice Hall, Englewood Cliffs (1993)
12. Beran, R., Bilodeau, M., de Micheaux, P.L.: Nonparametric tests of independence between random vectors. J. Multivar. Anal. **98**(9), 1805–1824 (2007)
13. Box, G.E.P., Pierce, D.A.: Distribution of residual correlations in autoregressive-integrated moving average time series models. J. Am. Stat. Assoc. **65**, 1509–1526 (1970)
14. Breusch, T.S., Pagan, A.R.: A simple test for heteroscedasticity and random coefficient variation. Econometrica **47**(5), 1287–1294 (1979)
15. Brodsky, B., Darkhovsky, B.: Nonparametric Methods in Change—Point Problems. Springer, New York (1993)
16. Brown, M.B., Forsythe, A.B.: Robust tests for equality of variances. J. Am. Stat. Assoc. **69**, 364–367 (1974)

17. Buishand, T.A.: Some methods for testing the homogeneity of rainfall records. J. Hydrol. **58**, 11–27 (1982)
18. Buishand, T.A.: Tests for detecting a shift in the mean of hydrological time series. J. Hydrol. **63**, 51–69 (1984)
19. Carlin, B.P., Gelfand, A.E., Smith, A.F.M.: Hierarchical Bayesian analysis of change-point problems. Appl. Stat. **41**, 389–405 (1992)
20. Chandola, V., Banerjee, A., Kumar, V.: Anomaly detection: a survey, ACM Comput. Surv. **41**(3), 15:1–15:58 (2009)
21. Chen, J., Gupta, A.: Parametric Statistical Change Point Analysis. Birkhauser Verlag, Basel (2000)
22. Chernoff, H., Zacks, S.: Estimating the current mean of a normal distribution which is subjected to change in time. Ann. Math. Stat. **35**, 999–1018 (1964)
23. Cochran, W.G.: The distribution of the largest of a set of estimated variances as a fraction of their total. Ann. Human Genet. (London) **11**(1), 47–52 (1941)
24. Cochrane, D., Orcutt, G.H.: Application of least squares regression to relationships containing auto-correlated error terms. J. Am. Stat. Assoc. **44**(245), 32–61 (1949)
25. Conover, W.J.: Practical Nonparametric Statistics. Wiley, New York (1971)
26. Csorgo, M., Horvath, L.: Limit Theorems in Change-Point Analysis. Wiley, New York (1997)
27. David, H.A.: Upper 5 and 1 % points of the maximum F-ratio. Biometrika **38**, 422–424 (1952)
28. Delgado, M.A.: Testing the serial independence using the sample distribution function. J. Time Ser. Anal. **17**, 271–285 (1996)
29. Diks, C.: Nonparametric tests for independence. http://www1.fee.uva.nl/cendef/upload/6/ecss_diks_r1.pdf
30. Dixon, W.J.: Analysis of extreme values. Ann. Math. Stat. **21**(2), 488–506 (1950)
31. Duggins, J., Williams, M., Kim, D.-Y., Smith, E.: Change point detection in SPI transition probabilities. J. Hydrol. **388**(3–4), 456–463 (2010)
32. Durbin, J.: Testing for serial correlation in least squares regression when some of the regressors are lagged dependent variables. Econometrica **38**, 410–421 (1970)
33. Durbin, J., Watson, G.S.: Testing for serial correlation in least squares regression I. Biometrika **37**, 409–429 (1950)
34. Durbin, J., Watson, G.S.: Testing for serial correlation in least squares regression II. Biometrika **38**, 159–178 (1951)
35. Durbin, J., Watson, G.S.: Testing for serial correlation in least squares regression III. Biometrika **71**, 1–19 (1971)
36. Embrechts, M., Maejima, P.: Self—Similar Processes. Princeton University Press, Princeton (2002)
37. Erdman, C., Emerson, J.W.: Bcp: an R package for performing a Bayesian analysis of change point problems. J. Stat. Softw. **23**(3), 1–13 (2007)
38. Erramilli, A., Narayan, O., Willinger, W.: Experimental queueing analysis with long—range dependent packet traffic. IEEE/ACM Trans. Netw. **4**(2), 209–223 (1996)
39. García, J.E., González-López, V.A.: Independence tests for continuous random variables based on the longest increasing subsequence. J. Multivar. Anal. **127**, 126–146 (2013)
40. Gedikli, A., Aksoy, H., Unal, N.E.: AUG—segmenter: a user-friendly tool for segmentation of long time series. J. Hydroinform. **12**(3), 318–328 (2010)
41. Gedikli, A., Aksoy, H., Unal, N.E., Kehagias, A.: Modified dynamic programming approach for offline segmentation of long hydrometeorological time series. Stoch. Env. Res. Risk Assess. **24**(5), 547–557 (2010)
42. Genest, C., Nešlehová, J.G., Rémillard, B.: On the estimation of Spearman's rho and related tests of independence for possibly discontinuous multivariate data. J. Multivar. Anal. **117**, 214–228 (2013)
43. Geweke, J., Porter-Hudak, S.: The estimation and application of long memory time series models. J. Time Ser. Anal. **4**, 221–238 (1983)

44. Glejser, H.: A new test for heteroscedasticity. J. Am. Stat. Assoc. **64**(325), 316–323 (1969)
45. Goldfeld, S.M., Quandt, R.E.: Some tests for homoscedasticity. J. Am. Stat. Assoc. **60**(310), 539–547 (1965)
46. Grubbs, F.E.: Procedures for detecting outlying observations in samples. Technometrics **11** (1), 1–21 (1969)
47. Hackl, P., Maderbacher, M.: On the robustness of the rank-based CUSUM chart against autocorrelation (1999). http://epub.wu.ac.at/1764/1/document.pdf
48. Hartley, H.O.: The maximum F-ratio as a short cut test for heterogeneity of variance. Biometrika **37**, 308–312 (1950)
49. Hawkins, D.M.: Identification of Outliers. Chapman and Hall, London (1980)
50. Higuchi, T.: Approach to an irregular time series on the basis of the fractal theory. Physica D **31**, 277–283 (1988)
51. Hines, W.G.S., Hines, R.J.O.: Increased power with modified forms of the Levene (med) test for heterogeneity of variance. Biometrics **56**(2), 451–454 (2000)
52. Hodge, V.J., Austin, J.: A survey of outlier detection methodologies. Artif. Intell. Rev. **22**(2), 85–126 (2004)
53. http://cran.r-project.org/web/packages/cpm/cpm.pdf
54. http://cran.r-project.org/web/packages/fArma/fArma.pdf
55. http://cran.r-project.org/web/packages/fBasics/fBasics.pdf
56. http://cran.r-project.org/web/packages/fractal/fractal.pdf
57. http://cran.r-project.org/web/packages/ecp/ecp.pdf
58. http://cran.r-project.org/web/packages/energy/energy.pdf
59. http://cran.r-project.org/web/packages/het.test/het.test.pdf
60. http://www.inside-r.org/packages/cran/lawstat/docs/levene.test
61. http://cran.r-project.org/web/packages/lmtest/lmtest.pdf
62. http://cran.r-project.org/web/packages/nortest/nortest.pdf
63. http://cran.r-project.org/web/packages/outliers/outliers.pdf
64. http://cran.r-project.org/web/packages/portes/portes.pdf
65. http://cran.r-project.org/web/packages/pracma/pracma.pdf
66. http://www.itl.nist.gov/div898/handbook/eda/section3/eda35h2.r
67. http://www.itl.nist.gov/div898/handbook/eda/section3/eda35h3.r
68. Hubert, P.: The segmentation procedure as a tool for discrete modeling of hydrometeorogical regimes. Stoch. Environ. Res. Risk Assess. **14**, 297–304 (2000)
69. Hubert, P., Carbonnel, J.P., Chaouche, A.: Segmentation des séries hydrométéorologiques. Application à des séries de précipitations et de débits de l' Afrique de l'Ouest, Journal of Hydrology **110**, 349–367 (1989)
70. Hurst, H.E.: Long-term storage of reservoirs: an experimental study. Trans. Am. Soc. Civ. Eng. **116**, 770–799 (1951)
71. James, N.A., Matteson, D.S.: Ecp: an R package for nonparametric multiple change point analysis of multivariate data. J. Stat. Softw. **62**(7), 1–25. http://www.jstatsoft.org/v62/i07/ paper
72. Jarque, C.M., Bera, A.K.: Efficient tests for normality, homoscedasticity and serial independence of regression residuals. Econ. Lett. **6**(3), 255–259 (1980)
73. Karagiannis, T., Faloutsos, M.: SELFIS: a tool for self—similarity and long—range dependence analysis (2002). http://alumni.cs.ucr.edu/~tkarag/papers/kdd02.pdf
74. Karagiannis, T., Faloutsos, M., Molle, M.: A user-friendly self-similarity analysis tool. ACM SIGCOMM Comput. Commun. Rev. Spec. Sect. Tools Technol. Netw. Res. Educ. **33** (3), 81–93 (2003)
75. Kehagias, A., Nidelkou, E., Petridis, V.: A dynamic programming segmentation procedure for hydrological and environmental time series. Stoch. Env. Res. Risk Assess. **20**(1–2), 77–94 (2006)
76. Keogh, E., Kasetty, E.: On the need for time series data mining benchmarks: A survey and empirical demonstration. Data Min. Knowl. Disc. **7**(4), 349–371 (2003)

77. Killick, R., Fearnhead, P., Eckley, I.A.: Optimal detection of changepoints with a linear computational cost. J. Am. Stat. Assoc. **107**(500), 1590–1598 (2012)
78. Kolmogorov, A.: Sulla determinazione empirica di una legge di distribuzione. Giornale dell'Istituto Italiano degli Attuari **4**, 83–91 (1933)
79. Lee, A.F.S., Heghinian, S.M.: A shift of mean level in a sequence of independent normal random-variables—Bayesian-approach. Technometrics **19**, 503–506 (1997)
80. Levene, H.: Robust tests for equality of variances. In: Olkin, I., Hotelling, H. et al. (eds.) Contributions to Probability and Statistics: Essays in Honor of Harold Hotelling, Stanford University Press, Stanford, pp. 278–292 (1960)
81. Li, W., McLeod, A.: Distribution of the residual autocorrelations in multivariate ARMA time series models. J. Roy. Stat. Soc. B **43**, 231–239 (1981)
82. Lilliefors, H.: On the Kolmogorov-Smirnov test for normality with mean and variance unknown. J. Am. Stat. Assoc. **62**, 399–402 (1967)
83. Lin, J., Keogh, E., Lonardi, S., Chiu, B.: A symbolic representation of time series, with implications for streaming algorithms, In: DMKD '03: Proceedings of the 8th ACM SIGMOD Workshop on Research Issues in Data Mining and Knowledge Discovery, pp. 2–11 (2003)
84. Liu, P., Guo, S., Xiong, L., Chen, L.: Flood season segmentation based on the probability change-point analysis technique. Hydrol. Sci. J. **55**(4), 540–554 (2010)
85. Ljung, G.M., Box, G.E.P.: On a measure of lack of fit in time series models. Biometrika **65**, 297–303 (1978)
86. Lo, A.W.: Long term memory in stock market prices. Econometrica **59**, 1279–1313 (1991)
87. Maddala, G.S., Lahiri, K.: Introduction to Econometrics. Wiley, Chichester (2009)
88. Mandelbrot, B.B., Wallis, J.R.: Robustness of the rescaled range R/S in the measurement of noncyclic long-run statistical dependence. Water Resour. **5**, 967–988 (1969)
89. Matilla-García, M., Rodriguez, J.M., Marin, M.R.: A symbolic test for testing independence between time series. J. Time Series Anal. **31**, 76–85 (2010)
90. Monti, A.C.: A proposal for residual autocorrelation test in linear models. Biometrika **81**, 776–780 (1994)
91. Moore, D.S.: Tests of the chi-squared type. In: D'Agostino, R.B., Stephens, M.A. (eds.) Goodness-of-Fit Techniques. Marcel Dekker, New York (1986)
92. Nelson, L.S.: Upper 10 %, 5 % and 1 % points of the maximum F-ratio. J. Qual. Technol. **19**(3), 165–167 (1987)
93. Neyman, J., Pearson, E.S.: On the use and interpretation of certain test criteria. Biometrika **20**, 175–240 (1928)
94. Noguchi, K., Gel, Y.R., Combination of Levene-type tests and a finite-intersection method for testing equality of variances against ordered alternatives. Working paper, Department of Statistics and Actuarial Science, University of Waterloo (2009)
95. O'Brien, R.G.: Robust techniques for testing heterogeneity of variance effects in factorial designs. Psychometrika **43**, 327–344 (1978)
96. Pearson, K.: On the criterion that a given system of deviations from the probable in the case of a correlated system of variables is such that it can be reasonably supposed to have arisen from random sampling. Phil. Mag. **5**, 157–175 (1900)
97. Peng, C.K., Mietus, J., Hausdorff, J.M., Havlin, S., Stanley, H.E., Goldberger, A.: Long-range anti-correlations and non-Gaussian behavior of the heartbeat. Phys. Rev. Lett. **70**, 1343–1346 (1993)
98. Peng, C., Buldyrev, S., Havlin, S., Simons, M., Stanley, H., Goldberger, A.: Mosaic organization of DNA nucleotides. Phys. Rev. E **49**(2), 1685–1689 (1994)
99. Pettitt, A.N.: A non—parametric approach to the change—point problem. Appl. Stat. **28**(2), 126–135 (1979)
100. Pinkse, J.: A consistent nonparametric test for serial independence. J. Econ. **84**, 205–231
101. Ray, B., Tsay, R.: Bayesian methods for change-point detection in long-range dependent processes. J. Time Ser. Anal. **23**, 687–705 (2002)

102. Razali, N.M., Wah, B.Y.: Tests for normality: comparison of powers, tests for normality: comparison of powers. J. Stat. Model. **2**(1), 2 1–33 (2011)
103. Robinson, P.M.: Consistent nonparametric entropy-based testing. Rev. Econ. Stud. **58**, 437–453 (1991)
104. Robinson, P.M.: Gaussian semiparametric estimation of long-range dependence. Ann. Stat. **23**, 1630–1661 (1995)
105. Robson, A. (2000) Analysis guidelines. In: Kundzewicz, Z.W., Robsson, A. (eds.) Detecting Trend and Other Changes in Hydrological Data, WCDMP-45, WMO/TD-No. 1013, pp. 11–14
106. Rosner, B.: Percentage points for a generalized ESD many-outlier procedure. Technome-trics **25**(2), 165–172 (1983)
107. Ross, G.J.: Parametric and Nonparametric Sequential Change Detection in R: The cpm package, www.gordonjross.co.uk/cpm.pdf
108. Scheffé, H.: The Analysis of Variance. Wiley, New York (1959)
109. Scott, A.J., Knott, M.: A cluster analysis method for grouping means in the analysis of variance. Biometrics **30**(3), 507–512 (1974)
110. Shapiro, S.S., Wilk, M. B.: An analysis of variance test for normality (complete samples). Biometrika **52**(3/4), 591–611 (1965)
111. Smirnov, N.: Table for estimating the goodness of fit of empirical distributions. Ann. Math. Stat. **19**, 279–281 (1948)
112. Steele, M., Chaseling, J.: Goodness-of-fit tests powers of discrete goodness-of-fit test statistics for a uniform null against a selection of alternative distributions. Commun. Stat. Simul. Comput. **35**, 1067–1075 (2006)
113. Stephens, M.A.: Tests based on EDF statistics. In: D'Agostino, R.B., Stephens, M.A. (eds.) Goodness-of-Fit Techniques. Marcel Dekker, New York (1986)
114. Szekely, G.J., Rizzo, M.L.: The distance correlation t-test of independence in high dimension. J. Multiv. Anal. **117**, 193–213 (2013)
115. Szekely, G.J., Rizzo, M.L., Bakirov, N.K.: Measuring and testing dependence by correlation of distances. Ann. Stat. **35**(6), 2769–2794 (2007)
116. Taqqu, M.S., Teverovsky, V.: Testing for long-range dependence in the presence of shifting means or a slowly declining trend, using a variance-type estimator. J. Time Ser. Anal. **18**(3), 279–304 (1997)
117. Taqqu, M.S., Teverovsky, V., Willinger, W.: Estimators for long-range dependence: an empirical study. Fractals **3**(4), 785–798 (1995)
118. Tsakalias, G., Koutsoyiannis, D.: A comprehensive system for the exploration and analysis of hydrological data. Water Resour. Manage **13**(4), 269–302 (1999)
119. Wang, G., Zou, C., Wang, Z.: Necessary test for complete independence in high dimensions using rank-correlations. J. Multivar. Anal. **121**, 224–232 (2013)
120. Warrenliao, T.: Clustering of time series data—a survey Pattern Recogn. **38**(11), 1857–1874 (2005)
121. Weron, R.: Estimating long range dependence: finite sample properties and confidence intervals. Phys. A **312**, 285–299 (2002)
122. White, H.: A heteroskedasticity-consistent covariance matrix estimator and a direct test for heteroskedasticity. Econometrica **48**(4), 817–838 (1980)
123. Zhang, Y., Meratnia, N., Havinga, P. Outlier detection techniques for wireless sensor networks: a survey. IEEE Commun. Surv. Tutorials **12**(2), 1–12 (2010)

Chapter 2
Mathematical Methods Applied for Hydro-meteorological Time Series Modeling

This chapter contains an overview of some methods that can be used for modeling the evolution of hydro-meteorological time series, methods employed in the next chapters. We distinguish among them: decomposition methods, Box–Jenkins methods and artificial intelligence—based methods (GEP, AdaGEP, GRNN and SVM).

For clarity sake, we remember some basic notions.

A time series is a set of successive observations of a phenomenon during a period [2]. A model is a formalized presentation of a phenomenon using an equation or a set of equations with the aim of understanding and explaining its evolution [14]. Time series modeling is the ensemble of techniques used for identifying a mathematical model representing a time series [29].

If (X_t) is a sequence of random variables and (x_t) is a sequence of the realizations of (X_t), building a time series model for (x_t) means the detection of the joint distribution of (X_t) [16].

In the following the notion of time series will be used both in the sense of [2] and [16].

Modeling and predicting nature phenomena and processes is a difficult task due to their variability that is a consequence of their relations with other phenomena and to the influences of unpredictable factors. Hydrological series are not an exception from this rule, their evolution being usually described by stochastic processes.

Two big groups of models for time series evolution can be determined: the classical one, containing the deterministic models, and the modern ones, composed of the models issued from the artificial intelligence, as evolutionary computing (EC) and artificial neural networks (ANNs) [32].

From other viewpoint, the modeling methods could be classified as parametric, nonparametric and semiparametric. Usually, in the parametric approach there are restrictive assumptions on the data, while, in the nonparametric one, the hypotheses on the input data are relaxed. Recent studies provided good results of modeling hydro-meteorological time series using ANNs and Gene Expression Programming (GEP) [6, 7, 23, 31, 34, 35]. Support Vector Regression (SVR), the adaptation of Support Vector Machines (SVM) for solving classification and regression

© Springer International Publishing Switzerland 2016   51
A. Bărbulescu, *Studies on Time Series Applications in Environmental Sciences*,
Intelligent Systems Reference Library 103, DOI 10.1007/978-3-319-30436-6_2

problems, proved to be a good alternative to the classical methods, with applications in different domains [12, 18, 45, 62]. Also, AdaGEP, a version of GEP, was successfully used for modeling hydrological time series, as an alternative to the Box–Jenkins methods [7, 8] or together with them, in hybrid models.

Due to the temporal and spatial variability of the hydro-meteorological time series, the use of only one method is not appropriate. Therefore, different techniques are employed; they are shortly presented in this chapter.

1   Types of Models. Classical Decomposition Method

The time series, in general, and the hydrological ones, in particular, may present trend and seasonality and are influenced by random variations. The existence of such components is differently considered, depending on the type of models.

A classification of the models is provided here [4]:

- Adjustment models, written as:

$$y_t = f(t, u_t),$$

where: f is an adjustment function, t is the time and u_t is a centered random variable.

The adjustment can be deterministic or random, as all components involved in the model are deterministic, or there is at least a random component.

The most common models with random adjustment are:

(a) the additive model, whose equation is:

$$y_t = Y_t + S_t + \varepsilon_t, \tag{1}$$

where: Y_t is the trend, S_t is the seasonal compound and ε_t is the random variable.

Taking into account the seasonality presence, (1) can be written as:

$$y_{ij} = Y_{ij} + S_j + \varepsilon_{ij}, \tag{2}$$

where: $i = \overline{1, p}$ is the period number, $j = \overline{1, m}$ is the number of the sub-period, Y_{ij} is the trend, S_j is the seasonality index and ε_{ij} is the random variable.

(b) the multiplicative model, whose equation is:

$$y_{ij} = Y_{ij} \cdot S_j^* \cdot \varepsilon_{ij}^*, \tag{3}$$

where: Y_{ij} is the trend, S_j^* is the seasonality index and ε_{ij}^* is the random variable.

- The autoprojective models, of the type:

$$y_t = f(y_{t-1}, y_{t-2}, \ldots, \varepsilon_t),$$

where ε_t is a random variable.
A special class of this type is formed by ARIMA models that will be presented in the next section.
- Explicative models, of the type:

$$y_t = f(X_t, u_t),$$

where X_t is an exogenous deterministic or random variable and u_t is an exogenous random variable.
In this case, X_t and u_t have some properties of independence or are not correlated.

Suppose that, for the moment, we analyze the additive model (2). Detecting each component is done by the classical method that supposes the steps:

1. Determining the trend, Y_{ij}, by the moving average method or an analytical method;
2. Determining the seasonal component as follows:

   (a) Determining the differences:

$$y_{ij} - Y_{ij} = S_j + \varepsilon_{ij}. \tag{4}$$

   (b) Determining the brute estimators of the seasonal component, s'_j, as averages of the differences (4) corresponding to the same season. If the sum of the brute estimators is not zero, pass to the step (c). If it is zero, pass to (d).
   (c) Determining the seasonal component, s_j, by subtracting the average of the brute estimators of the seasonal components from the brute estimator.
   (d) Supposing that $s_j = S_j$, determine the values of the random variable by:

$$\varepsilon_{ij} = y_{ij} - Y_{ij} - S_j. \tag{5}$$

For the multiplicative model (3), the decomposition is done by an analogous algorithm, whose steps are:

1. Determining the trend, Y_{ij};
2. Determining the seasonal component as follows:

   (a) Determining the ratio:

$$y_{ij}/Y_{ij} = S_j^* \cdot \varepsilon_{ij}^*, \tag{6}$$

   (b) Determining the brute estimators of the seasonal component, s_j'', as averages of the ratios (6) corresponding to the same season. If the product of the brute estimators is not one, pass to the step c. If it is one, pass to d.

(c) Determining the seasonal component, s_j^*, by dividing the brute estimators by the average of the brute estimators of the seasonal components;

(d) Supposing that $s_j^* = S_j^*$, determine the values of the random variable by:

$$\varepsilon_{ij}^* = y_{ij}/(Y_{ij} \cdot S_j^*).$$

To perform such a decomposition, 'decompose()' function in R may be used. It estimates the components of a time series that is described by an additive model. The function has the following arguments:

- x—the time series that will be decomposed;
- type—type of model—"additive" of "multiplicative"; the default is the additive one;
- filter—a filter coefficient—if "NULL", a moving average fit is performed.

The following sequence of code must be written for the decomposition of Constanta monthly series, by the multiplicative model (Fig. 1).

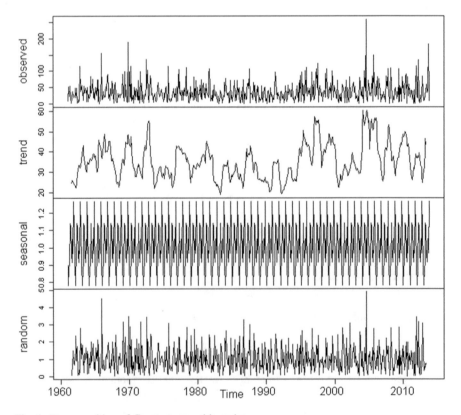

Fig. 1 Decomposition of Constanta monthly series

```
data<-read.csv("D:\\Lucrari_2.12.14\\2015_Carte\\Cta_lunar_1961_2013.csv",sep=",",
header=TRUE)
  x <- data[,1]
  library(stats)
  x<-ts(x, start=c(1961,1), end=c(2013,12), frequency =12)
  m <- decompose(x,type="multiplicative)
```

```
    [1] 0.9005163 0.7813944 0.8740163 0.9073708 1.1396821 1.1196095 0.9607246
    [8] 0.9139141 1.0496215 0.9650946 1.2695325 1.1185233
    # gives seasonal figures
```

2 Box–Jenkins Approach and Stationarity Tests

2.1 Box–Jenkins Approach

Box and Jenkins proposed in 1970 [15] a general approach for modeling and predicting univariate time series, based on the notion of ARMA process, that can be successfully used for modeling series that don't present very high variability. In the following we summarize the basic notions [16].

The process (X_t) is said to be *weakly stationary* (stationary, for short) if it has a constant mean and its covariance depends only on the lag between two points in the series.

A process (X_t) is called *white noise* if (X_t) is uncorrelated, identically distributed, with zero mean, and a constant variance.

A discrete process (X_t) is called of AR(p) type (*autoregressive of p order*) if it can be described by the equation:

$$X_t = c + \sum_{i=1}^{p} \varphi_i X_{t-i} + \varepsilon_t, \quad \varphi_p \neq 0, \tag{7}$$

where c is a constant, $\varphi_i, i = \overline{1, p}$ are parameters and (ε_t) is a white noise.

It was proved that a sufficient condition of stationarity of an AR(p) process is that the absolute values of all the roots of the characteristic equation:

$$z^p - \sum_{i=1}^{p} \varphi_i z^{p-i} = 0, \tag{8}$$

are less than 1.

Particularly, if $|\varphi_1| > 1$, the AR(1) process is not stationary.

A discrete process (X_t) is called of MA(q) type (*moving average of q order*) if it has the equation:

$$X_t = \mu + \varepsilon_t - \sum_{i=1}^{q} \theta_i \varepsilon_{t-i}, \quad \theta_q \neq 0, \tag{9}$$

where μ is a constant, $\theta_i, i = \overline{1, q}$ are parameters and (ε_t) is a white noise.

It was proved that a sufficient condition for the stationarity of an MA(q) process is that the absolute values of the roots of the characteristic equation

$$z^q - \sum_{i=1}^{p} \theta_j z^{q-j} = 0 \tag{10}$$

are less than 1.

If (X_t) is a discrete process, *the partial autocorrelation function* (PACF) is defined by:

$$\tau(h) = \frac{\mathrm{Cov}(X_t - X_t^*, X_{t-h} - X_{t-h}^*)}{\left[\mathrm{D}^2(X_t - X_t^*) \mathrm{D}^2(X_{t-h} - X_{t-h}^*) \right]^{1/2}}, \quad t > h \geq 0,$$

where $X_t^*(X_{t-h}^*)$ is the affine regression of $X_t(X_{t-h})$ on $X_{t-1}, \ldots, X_{t-h+1}$.

The autocorrelation and the partial autocorrelation functions can be used for the selection of the model's type. The partial autocorrelation function of an AR(p) model is zero for the lag h greater than p, and its ACF is a sum of decaying exponentials or a damping sine wave. The autocorrelation function of an MA(q) model is zero for the lag h greater than q.

A process (X_t) is called of ARMA(p, q) type (*autoregressive moving average process of autoregressive order p and moving average order q*) if it can be described by the equation:

$$X_t - \sum_{i=1}^{p} \varphi_i X_{t-i} = \theta_0 + \varepsilon_t - \sum_{i=1}^{q} \theta_i \varepsilon_{t-i}, \quad \varphi_p \neq 0, \ \theta_q \neq 0, \tag{11}$$

where θ_0 is a constant, $\varphi_i, i = \overline{1, p}, \theta_i, i = \overline{1, q}$ are parameters and (ε_t) is a white noise.

If B is the backward operator, defined by:

$$B^k(X_t) = X_{t-k}, \quad t > k, t, k \in \mathbf{N}^*,$$

and:

$$\Phi(z) = 1 - \varphi_1 z - \varphi_2 z^2 - \cdots - \varphi_p z^p, \quad \varphi_p \neq 0,$$
$$\Theta(z) = 1 - \theta_1 z - \theta_2 z^2 - \cdots - \theta_q z^q, \quad \theta_q \neq 0,$$

the relations (7), (9), (11) can be written respectively as:

$$\Phi(B)X_t = c + \varepsilon_t,$$
$$X_t = \mu + \Theta(B)\varepsilon_t,$$
$$\Phi(B)X_t = \theta_0 + \Theta(B)\varepsilon_t.$$

AR(p) and MA(q) are particular cases of ARMA(p, q) processes: AR (p) = ARMA(p, 0), MA(q) = ARMA(0, q).

Preliminary estimation of the parameters in AR models can be done by Yule-Walker and Burg [17] procedures, and in ARMA models, by the innovation and Hannan and Rissanen [30] algorithms.

For selecting the best ARMA(p, q) model that fit the data, the Akaike [1] and Schwarz [57] criteria are used. These are defined by:

$$AIC(p,q) = \ln \hat{\sigma}^2_{p,q} + 2(p+q)/n,$$
$$SCH(p,q) = \ln \hat{\sigma}^2_{p,q} + (p+q)\ln n/n,$$

$\hat{\sigma}^2_{p,q}$ being the maximum likelihood estimation of the errors' variance and n, the volume of the sample used to fit the model.

The best ARMA(p, q) model is that with the smallest AIC (or SCH) value.

A process (X_t) is called of ARIMA(p, d, q) type (*autoregressive integrated moving average process*) if it can be described by the equation:

$$\Phi(B)(1 - B)^d X_t = \Theta(B)\varepsilon_t,$$

where $d \in \mathbf{N}^*$ is the degree of differentiation, (ε_t) is a white noise and $\Phi(z) \neq 0$, for all z with $|z| \leq 1$.

A necessary and sufficient condition for the stationarity of an ARIMA(p, d, q) process is $d = 0$, in which case this process reduces to an ARMA(p, q) process.

A discrete process is said to be a FARIMA(p, d, q) (*fractionally integrated ARMA*) process if:

$$(1 - B)^d \Phi(B)X_t = \Theta(B)\varepsilon_t,$$

where $\Phi(z) \neq 0, \Theta(z) \neq 0$ for all z with $|z| \leq 1, (\varepsilon_t)$ is a white noise, d is the memory parameter, $0 < |d| < 0.5$ and

$$(1 - B)^d = \sum_{j=0}^{\infty} \pi_j B^j,$$

with

$$\pi_j = \prod_{1 < k \leq j} \frac{k - 1 - d}{k}, \quad j = 1, 2, \ldots \text{and } \pi_0 = 1.$$

For an extensive study of the Box–Jenkins approach, see [15].

The R stats package [40] can be used to simulate an ARIMA model. Here are some examples:

```
library(stats)
ts.sim<-arima.sim(n=40, list(ar=c(0.451, -0.485), ma=c(-0.879, 0.748)), sd=sqrt(0.234))
#generates 40 values of an ARMA(2,2) model, without constant term, whose autoregressive
# and moving average coefficients are 0.451 and -0.485, respectively 0.879 and 0.748
ts.sim #print the simulated values
```

```
Time Series:
Start = 1
End = 40
Frequency = 1
```
[1]   0.45851204 0.29764486 -0.48782867 1.00328887 0.23271658 -0.02816671
 [7]   0.22655481 0.19091807 -0.35784716 -0.88659972 1.59331365 -0.45864452
[13] -0.36238013 1.40164581 -0.54335712 0.38079291 0.34834893 -0.69216474
[19] 0.41504097 0.22675742 -0.20430218 -0.17780706 0.14819071 0.60619994
[25] -0.65479046 -0.49896103 -0.04767047 -0.19592509 -0.98984893 0.73034419
[31] -0.58868968 0.41709830 0.52431536 0.14624364 -0.67246452 0.73284864
[37] -0.10253791 0.23828534 0.69601486 -0.32475338

```
ts.plot(ts.sim)#plots the generated data series (Fig. 2)
ts.sim <- arima.sim(list(order=c(2, 1, 0), ar= c(0.2, 0.45)), n=20)
#generates an ARIMA(2,1,0) model
ts.sim
```

```
Time Series:
Start = 1
End = 21
Frequency = 1
```
[1]   0.0000000 0.5968887 2.0204269 2.2838219 2.8037469 3.2241359
 [7]   3.5735482 4.0904095 4.5394847 5.0389145 5.8879888 5.2578066
[13] 3.7503221 4.1217022 2.7240303 3.0765823 1.3698995 2.0259479
[19] 0.2043944 0.9807588 -0.8276302

```
ts.plot(ts.sim) (Fig. 3)
ts.sim <- arima.sim(list(order = c(0,0,0)), n = 50) # generates a white noise
# the same can be done by using 'rnorm(50)' (Fig. 4)
```

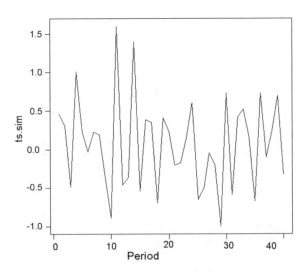

Fig. 2 Simulation of an ARMA(2, 2) model

Fig. 3 Simulation of an
ARIMA(2, 1, 0) model

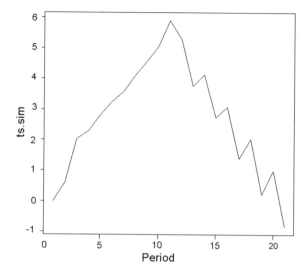

Fig. 4 Simulation of a white
noise

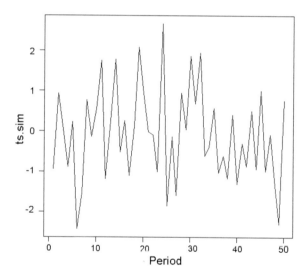

Different ways can be followed for fitting an ARIMA model. The first one is to choose own model(s) using Arima() function from the R package **forecast** [36]. Among the models that fit the data, the selected one is that with the smallest AIC.

Another way is to utilize auto.arima() [43] which is based on the iterative algorithm introduced by Hyndman and Khandakar [44] and that combines the KPSS stationarity test (for determination of differences order d) and AIC computation (for choosing the parameters p and q for the differentiated series). If $d = 0$, a constant is also included in the model.

We exemplify here the use of auto.arima() on Constanta daily precipitation series recorded over a period of 10 years.

```
library(forecast)
data<-read.csv("D:\\Lucrari_2.12.14\\2015_Carte\\Cta_zilnice_1961_2009.csv", sep=",",
header=TRUE)
  y<-data[1 :3850,1] #selects only the first 3850 data on the first column of the sheet
  fit <- auto.arima(y) #fit the ARIMA model
  plot(forecast(fit, h = 120)) #plots the forecast for the next 120 days based on the model
# (Fig. 5)
  fit
```

```
        Series: y
        ARIMA(0,0,1) with non-zero mean

        Coefficients:
                ma1     intercept
              0.1825    1.1499
        s.e.  0.0162    0.0766

        sigma^2 estimated as 15.31:  log likelihood=-10158.42
        AIC=20322.85    BIC=20341.46 #BIC is the Bayesian Information Criteria
```

Remark The model is not satisfactory, so other methods must be employed.

Fig. 5 MA(1) model for the daily precipitation registered a Constanta (1961–1970)

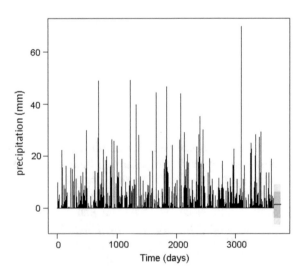

2.2 Stationarity Tests

Different statistical tests have been proposed for checking the hypothesis of time series stationarity. Some of them can be applied without knowledge on the model designed for the data series; others are used knowing the autoregressive models fitted on the data series, for testing the existence of a unit root. All can be grouped in two categories:

- Tests whose null hypothesis is that the series is non-stationary, as: the DF (Dickey and Fuller) [22, 25], ADF (Augmented Dickey–Fuller), PP (Phillips–Perron) [54] and SP tests (Schmidt and Phillips) [56] tests;
- Tests whose null hypothesis is that the series is stationary, as the KPSS (the Kwiatkowski, Phillips, Schmidt and Shin) test [47].

Generally, a time series (X_t) may have a stochastic trend (unit root) (TS_t), a deterministic trend and/or a cyclical component (C_t), such that it could be written as:

$$X_t = TD_t + TS_t + C_t. \tag{12}$$

Therefore, for testing the hypothesis $TS_t \neq 0$, the first kind of tests is used, and for checking $TS_t = 0$, the second one is employed.

In the Dickey–Fuller test, the following models are investigated: (1) autoregressive of first order: (2) autoregressive of first order, with an average different from zero, (3) autoregressive with trend. They are described respectively by the equations:

$$\Delta X_t = \varphi X_{t-1} + \varepsilon_t, \tag{13}$$

$$\Delta X_t = \varphi X_{t-1} + \beta + \varepsilon_t, \tag{14}$$

$$\Delta X_t = \varphi X_{t-1} + \beta + \gamma t + \varepsilon_t, \tag{15}$$

where ε_t is an independent random variable and Δ is the backshift operator.

Therefore, the null hypothesis H_0: $\varphi = 0$ (or, equivalent, the existence of a unit root) is tested against the stationarity one, H_1: $\varphi < 0$.

The null hypothesis is rejected if the p-value calculated for the test is less than the established significance level (0.05, if no other specification is done) [25, 53].

ADF is a generalization of DF test, which relies on the model:

$$\Delta X_t = \varphi X_{t-1} + \sum_{j=1}^{p-1} \alpha_j \Delta X_{t-j} + \beta' D_t + \varepsilon_t,$$

where: ε_t is a white noise and D_t is a deterministic trend. The null hypothesis is, again, the existence of a unit root ($\varphi = 0$).

The test statistic is the usual t test. A value of this statistics greater than the critical value (from the table of the Student t statistics) corresponding to the significance level indicates that the null hypothesis cannot be rejected.

In the case of PP test, the series' heteroskedasticity is also considered, so that the series model is:

$$\Delta X_t = \phi X_{t-1} + \beta' D_t + \varepsilon_t,$$

where ε_t is stationary and may be heteroskedastic.

The test statistics is a modified version of the ADF test statistics, which corrects the autocorrelation or heteroskedasticity of ε_t [39].

Belonging to the second group of tests, KPSS tests the null hypothesis that the study series is stationary in mean or in trend (linear deterministic).

The series' model is:

$$X_t = \beta' D_t + r_t + u_t, \quad r_t = r_{t-1} + \varepsilon_t,$$

where D_t is the deterministic component (constant or linear trend), r_t is a random walk, ε_t is stationary, with the variance $\sigma_\varepsilon^2, u_t$ is stationary, but may be heteroskedastic.

The null hypothesis is equivalent with: $H_0 : \sigma_\varepsilon^2 = 0$ and the test statistics used to test it is [39, 53]:

$$KPSS = \left(T^{-2} \sum_{t=1}^{T} \hat{S}_t^2 \right) \Big/ \hat{\lambda}^2,$$

where $\hat{S}_t = \sum_{j=1}^{t} \hat{u}_j, \hat{u}_j$ is the residual in the regression model of X_t on D_t and $\hat{\lambda}^2$ is a consistent estimate of the variance parameter:

$$\lambda^2 = \lim_{T \to \infty} \sum_{t=1}^{T} E[T^{-1} S_T^2].$$

The null hypothesis is rejected at a significance level if the test statistic is greater than the critical value.

For a deeper insight on these tests, the reader may refer to [53].

In [21] some of these tests have been applied to analyze the stationarity of precipitation series from Benin. Here, we present the code in R for performing some tests and the results of their application on Constanta daily series (1961–2009).

```
library(tseries)
data<-read.csv("D:\\Lucrari_2.12.14\\2015_Carte\\Cta_zilnice_1961_2009.csv", sep=",",
header=TRUE)
  y<-data[,1]
  adf.test(y)
```

> Augmented Dickey-Fuller Test
> data: y
> Dickey-Fuller = -8.7952, Lag order = 8, p-value = 0.01
> alternative hypothesis: stationary

> Warning message:
> In adf.test(y): p-value smaller than printed p-value

PP.test(y)

> Phillips-Perron Unit Root Test

data: y
Dickey-Fuller = -24.8673, Truncation lag parameter = 6, p-value = 0.01

kpss.test(y, "Level")

> KPSS Test for Level Stationarity
> data: y
> KPSS Level = 0.873, Truncation lag parameter = 5, p-value = 0.05952

kpss.test(y, "Trend")

> KPSS Test for Trend Stationarity
> data: y
> KPSS Trend = 0.1382, Truncation lag parameter = 5, p-value = 0.06437.

The results of ADF and PP tests prove that the hypothesis that the series has a unit root can be rejected. KPSS test could not reject the hypothesis that the series is stationary in level and trend.

The same tests are implemented in the packages **fUnitRoots** [37] and **urca** [38]. In the following we present some examples of the use of these tests.

```
library(fUnitRoots)
adfTest(y) #computes test statistics and p values along the implementation of Traplletti
# [37] for unit  roots
```

> Title:
> Augmented Dickey-Fuller Test

> Test Results:

```
PARAMETER:
Lag Order: 1
STATISTIC:
Dickey-Fuller: -7.4657
P VALUE:
0.01
```

```
Warning message:
In adfTest(y) : p-value smaller than printed p-value
```

unitrootTest(y) #computes test statistics and p values along the implementation of
McKinnon [50] for unit roots

```
Title:
Augmented Dickey-Fuller Test
```

```
Test Results:
```

```
PARAMETER:
Lag Order: 1
STATISTIC:
DF: -7.4657
P VALUE:
t: 1.737e-12
n: 0.05854
```

```
library(urca)
data<-read.csv("D:\\Lucrari_2.12.14\\2015_Carte\\Cta_zilnice_1961_2009.csv", sep= ", ",
header=TRUE)
 y<-data[,1]
 z <- ur.pp(y, type="Z-tau", model="trend", lags="short") # performs the PP test
 summary(z)
```

Test regression with intercept and trend

Call:
lm(formula = y ~ y.l1 + trend) # builds the regression equation of y

Residuals:
Min	1Q	Median	3Q	Max
-38.785	-19.877	-6.802	12.143	221.550

Coefficients:
Estimates the regression coefficients, the values of t statistics and the p-values

	Estimate	Std. Error	t value	Pr(>|t|)	
(Intercept)	34.964872	1.833083	19.074	<2e-16	***
y.l1	0.010005	0.039808	0.251	0.8016	
trend	0.011200	0.006425	1.743	0.0818	.

Signif. codes: 0 '***' 0.001 '**' 0.01 '*' 0.05 '.' 0.1 ' ' 1
y.l1 is not significant, the trend coefficient is significant at 0.01 and the trend is
also significant.

Residual standard error: 29.6 on 632 degrees of freedom
Multiple R-squared: 0.005011, Adjusted R-squared: 0.001862
there is no correlation between the exogenous and endogenous variables
F-statistic: 1.591 on 2 and 632 DF, p-value: 0.2044
the model is not significant in its whole
Value of test-statistic, type: Z-tau is: -24.8674

	aux. Z statistics
Z-tau-mu	19.3696
Z-tau-beta	1.7425

Critical values for Z statistics:
                 1pct   5pct   10pct
critical values -3.977072 -3.419005 -3.131727
The value of Z-tau is less than the critical values, so that the null hypothesis can
be rejected.

x<-ur.sp(y, type="tau", pol.deg=1, signif=0.01) # performs the SP test
summary(x)

Call:
lm(formula = sp.data)

Residuals:
     MIn   1Q   Median   3Q   Max
 -38.785 -19.877 -6.802   12.143 221.550

Coefficients:
	Estimate	Std. Error	t value	Pr(>|t|)	
(Intercept)	31.397725	2.672152	11.750	<2e-16	***
y.lagged	0.010005	0.039808	0.251	0.8016	
trend.exp1	0.011200	0.006425	1.743	0.0818	.

Signif. codes: 0 '***' 0.001 '**' 0.01 '*' 0.05 '.' 0.1 ' ' 1
summary(x)

Residual standard error: 29.6 on 632 degrees of freedom
Multiple R-squared: 0.005011, Adjusted R-squared: 0.001862
F-statistic: 1.591 on 2 and 632 DF, p-value: 0.2044

Value of test-statistic is: -22.2254
Critical value for a significance level of 0.01 is: -3.58
The hypothesis that the series has a unit root is rejected.

3   Genetic Algorithms

Evolutionary Algorithms (EAs) are part of artificial intelligence techniques, inspired by the theory of natural evolution and selection of species. So, a population generates new populations of individuals and among them, the best individuals survive. In these algorithms, the individuals are candidate solutions of the problem at hand. The environment in which the individuals live is the search space. The evolution is produced by the application of different genetic operators, and the selection process is done by using the fitness function, based on the principle that the best individual in a population survives.

In biology, the genes of an organism encode the rules that describe how the organism is built up. At their turn, the genes are connected in chromosomes.

In Evolutionary Algorithms, the genotype defines the individuals' content and structure, and the phenotype refers to the behavior of an individual genotype.

Before running an evolutionary algorithm for solving a problem, the potential solutions must be encoded as bit strings, referred as chromosomes, in this context. The flowchart of an EA is:

1. Generate an initial population of candidate solutions (randomly).
2. Assign a fitness score to each chromosome accordingly to his quality for solving the problem.
3. Apply genetic operators to the population.
4. Check the stop criteria. If it is fulfilled, go to 5. If not, restart from 2.
5. Stop.

The individuals' selection can be done by different methods, as the fitness-proportionate selection or the tournament selection [10].

Genetic Algorithm (GA) and Genetic Programming (GP) belong to the family of EAs. In GA the genes are encoded by fixed length binary strings, and the chromosomes of the population have the same size.

For evolving the initial population, Holland [33] defined the one-point cross-over, as a main genetic operator, and the one-point mutation and the inversion, as secondary operators. GA mutation acts by flipping a bit in an individual. By crossover, two offspring are built from two individuals by swapping some segments of genetic code between them.

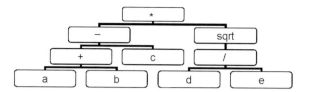

Fig. 6 GP individual that encodes the expression $[(a+b) - c] * \sqrt{d/e}$. The functions are the root node and inner nodes. The terminals are the leaf nodes

The main drawback of GA is the constant size of the chromosomes that does not permit the change of the model' structure during evolution. To eliminate this inconvenient, GP represents the individual using a tree structure, with a variable length. Figure 6 presents such an example. The trees can evolve, altering their shapes, sizes, and contents.

The function set can include algebraic operators, usual mathematical functions, Boolean operations, conditional operators (If—Then—Else), recursive functions, iterative functions (Do—Until). Terminals are variables or constants.

The symbols' set must be selected so as to meet the necessary and sufficient conditions to model the problem.

GP individuals encode computer programs, or mathematical functions expressed as complex compositions of functions and variables or constants [46]. If GAs are mainly used in optimization problems, GP is designed for models' identification, for given datasets. Contrary to what happens in nature, in GP there is no delimitation between the phenotype and the genotype. Therefore, the genetic operators act directly on the phenotype [7] and limit the operators' search power [24].

For an insight on this topic, one may refer to [46, 48].

3.1 Gene Expression Programming

Gene Expression Programming (GEP) has been introduced in 2001, by Candâda Ferreira [23]. It belongs to the family of GAs and is an automatic programming method based on the principle of natural selection. GEP combines futures of GA (linear chromosomes of fixed length) and GP (hierarchical structures of different forms and dimensions). Its basic idea is the representation of the solutions of the problem at hand as individuals that evolve over generations by means of genetic operators.

Our aim is to present here a short overview of GEP based on [6, 10, 23, 24], not to explore all its capabilities.

Unlike GP, that does not allow multiple genes per individual for building solutions, GEP individuals are formed by many genes of equal length, coding nonlinear expressions. The individuals are strings of symbols (mathematical functions, constants, or variables). At its turn, a GEP gene is formed by a head (that can contain constants, variables, and symbols) and a tail (that can contain only constants and variables). Every gene encodes a sub-tree. The sub-trees attached to every gene of an individual are linked together to the root node using linking functions that belong to a set specified by the user.

The gene number is a parameter that has to be set in the algorithm.

The proportionate roulette-wheel scheme [26] and simple elitism (cloning the best individual) are used for individuals' selection in GEP. The individuals of each generation are evaluated function of a performance measure. It is based on the error of the expression coded by the individual with respect to the input data. As a result, a fitness value is attached to each individual. The higher the fitness value is, the better an individual is. Therefore, the selection of the individuals for replication is done function of their fitness and the luck of the roulette. Then, the genomes of the best individuals (the selected ones) are copied as many times as the outcome of the roulette, during the replication process.

The operations involved in the genetic evolution are:

- Mutation, in which a part of an individual changes, preserving the chromosome' structural organization.
- Transposition (IS, RIS and gene transposition) that is used for copying a randomly selected part of a gene and moving it into another position of a chromosome. In the transposition process, a gene is deleted from its initial position and is moved at the beginning of the chromosome, such that the chromosome's length is maintained.
- Crossover, which combines futures of two parents in offspring. The GA crossover and its specific operators are also used in GEP, but a specific GEP operator for genes' crossover is also defined.

In the process of function finding, the goal is to determine an expression that links the endogenous variable to the exogenous ones. Particularly, in the time series modeling, given the sample $\{x_1, x_2, \ldots, x_n\}$, the aim is to determine a model that approximates as well as possible the given values.

Here are the steps in solving this type of problem:

- Define the fitness function for an individual program;
- Choose the terminals' set, T, and the functions' set, F, for creating the chromosomes. In the simplest case, $F = \{+, -, *, /\}$;
- Choose the chromosome' structure, i.e. the length of the head and the number of genes;
- Choose the linking function;
- Choose the set of genetic operators and their rates.

An important issue is the determination of the window size, w, i.e. the number of the previous values used to estimate the actual one.

Depending on w, the preprocessed time series provides n-dimensional elements, $d_{t-w} = (x_{t-w}, x_{t-w+1}, \ldots, x_{t-1})$, $w + 1 \leq t \leq n$, function of which x_t is estimated by GEP, as:

$$x_t = f(x_{t-w}, x_{t-w+1}, \ldots, x_{t-2}, x_{t-1}) + \varepsilon_t, \quad w + 1 \leq t \leq n.$$

In our experiments, the fitness is considered to be the mean squared error (MSE) of the model coded by the chromosomes with respect to the registered data, which is defined by:

$$MSE = \frac{1}{n-1} \sum_{t=1}^{n} (x_t - \hat{x}_t)^2,$$

where x_t is the registered value, $\hat{x}_t$ is the estimated one, and n is the sample's volume.

3.2 Adaptive Gene Expression Programming

In GEP the genes' number in a chromosome is constant for all individuals in a population. The modification of this number affects the length and the form of the solution.

Determining the optimum genes' number is a difficult, time-consuming process, involving the user's experience, requesting a big number of experiments. For avoiding this restriction, Adaptive Gene Expression Programming (AdaGEP) algorithm [13] is designed to identify automatically the genes' number in GEP, using a deactivation mechanism [7]. In this approach, a genemap is attached to each chromosome, in order to select the genes that participate in the decoding process. It is a bit string whose dimension is equal to the number of genes of a GEP individual. If the bit value is 1, the gene participates to the decoding, as in GEP case; if it is zero, it does not participate in this process.

The genemaps evolve similarly to the population in a GA; the mutation and crossover operations on them are exactly as in GA case. GEP iterations are accompanied by genemaps' iterations. Also, a genemap survives in the selection process only if the corresponding chromosome survives [7].

So, the steps of AdaGEP algorithm are:

1. Create the initial population of individuals (randomly)
2. Evolve individuals with GEP operators (crossover, mutation, transpositions)
3. Apply Gene Map Evolution operator on the population of genemaps
4. Evaluate each AdaGEP individual over the set of fitness cases
5. Select the next generation of individuals (by roulette wheel selection)

   - the fitness value of a genemap is that of the individual to whom it is attached.
   - survival of a GEP chromosome implies the survival of its genemap also.

6. Go to 2 if the stop criterion is not fulfilled.

4 Support Vector Regression (SVR)

Support Vector Regression (SVR) [55] is a nonlinear regression method derived from the Support Vector Machines technique (SVM) that was developed for solving problems of supervised learning and classification, based on the principle of errors' minimization. The idea behind this algorithm is the construction of a model function, f, which forecasts the output of a system that depends on a set of variables, using a set of input data for which the output is known.

The main characteristic of SVM is that the prediction function is developed on a support set. The algorithms based on Support Vector have been extended to classification problems, using different loss functions [3].

SVR is a category of SVM. ε-SVR, introduced by Vapnik [63] uses so-called ε-insensitive loss function, which minimizes the generalized error bound instead of minimizing the learning error, using the structural risk minimization principle. This algorithm searches a function f, as smooth as possible, that has at most a ε deviation from the specified output of all learning data. These constraints lead to a convex optimization problem.

Below we present the formulation of the algorithm, based on [3, 63].

Firstly, let us consider the case of a linear function defined by:

$$f(x) = \langle w, x \rangle + b, \quad b \in \mathbf{R}, \quad x \in X,$$

where X is the input space and $\langle ., . \rangle$ is the inner product in X.

Suppose that the learning data are denoted by $(x_i, y_i) \subset X \times \mathbf{R}, i = \overline{1, m}$.

For taking into account the existence of an infeasible convex optimization problem, the slack variables ξ, ξ^* are introduced, so that the problem becomes:

$$\text{minimize} \left\{ \frac{1}{2} \|w\|^2 + C \sum_{i=1}^{m} (\xi_i + \xi_i^*) \right\},$$

under the constraints:

$$\begin{cases} y_i - \langle w, x_i \rangle - b \leq \varepsilon + \xi_i \\ \langle w, x_i \rangle + b - y_i \leq \varepsilon + \xi_i^* \\ \xi_i, \xi_i^* \geq 0 \end{cases}$$

Using the dual problem, the function f can be written as:

$$f(x) = \sum_{i=1}^{m} (\alpha_i - \alpha_i^*) \langle x_i, x \rangle + b, \tag{12}$$

with

$$f(x) = \sum_{i=1}^{m} (\alpha_i - \alpha_i^*) = 0, \quad \alpha_i, \alpha_i^* \in [0, C].$$

Relation (12) is the support—vector expansion of f. C and ε are parameters that must be determined.

In v-SVR, which is a version of SVR, ε is considered to be a variable in the optimization process and a most convenient parameter $v \in (0, 1)$ is introduced.

For solving the non-linear problem, a projection of input data on a Hilbert space, F, of higher dimension is done and then (12) is written as:

$$f(x) = \sum_{i=1}^{m} (\alpha_i - \alpha_i^*) K(x_i, x) + b,$$

utilizing a kernel function, K, as linear, polynomial, radial basis function (RBF), sigmoid [35, 58] or other functions that fulfill Mercer's conditions [60].

Since there are no general criteria to choose a particular kernel, in most cases the choice is done in an empirical way.

The prediction requires the choice of kernel parameters, which depend on data. This is, generally, a difficult task. In many cases parameters are tuned by hand, based on experience, and are adjusted taking into account the experimental results.

Kernlab package from R can be used to run SVR. Other software may also be employed for this purpose. For some of them, the reader may refer to [41].

5 General Regression Neural Network (GRNN)

Artificial Neural Networks (ANNs) have been introduced by McCulloch and Pitts [49] for modeling logical functions. They are formed by artificial neurons that sum the input data producing the output. An activation function is applied to the weighted sums to produce the result, which becomes the input of the next layer.

General Regression Neural Networks (GRNN) were designed by Specht [59] as feed-forward networks, with four layers: Input layer, Pattern layer, Summation layer and Output layer (Fig. 7). Unlike Probabilistic Neural Networks that perform classifications on discrete variables, GRNN performs regressions and the target variables are continuous [60].

In GRNN, the numbers of neurons in the Input layer and that of the predictor variables are equal.

GRNN uses nonparametric estimators of the density of probability function. The measure of the estimation's quality realized by each training sample is the Euclidean or city-block distance between the training sample and the prediction point [59]. The distances between the assessed point and the other points are calculated and a kernel function is applied to them for computing the weight of each point. The best estimated value of the new point is determined by summing the

Fig. 7 GRNN diagram

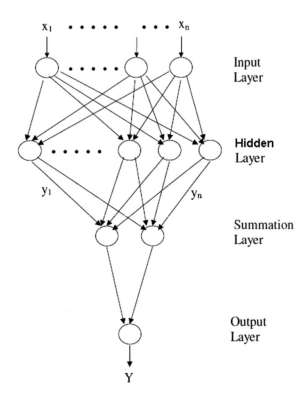

values of the other points, weighted by the kernel function. These operations are done in the Hidden layer, which is also used for storing the predictors' values and the target values and whose number of neurons is equal to the number of data in the training set.

The Summation layer is formed by two neurons: the numerator and the denominator summation neurons. They are used for the storage of the sum of the weights from the Hidden layer and the weights multiplied by the actual target values, respectively.

The Output layer is composed of one neuron that contains the result of the division of the values stocked by the numerator and the denominator of the previous layer [5].

6 Wavelets

Wavelets are waves with some specific properties, utilized for representing different functions [52], based on the wavelets transform (WT), a recent technique developed by Grossman and Morlet [28]. Due to its advantage, as the absence of the restrictions related to the process' stationarity or its long range dependence property

[27], and the preservation of the local phenomena, multiscale and periodical features, WT is a valuable tool for time series modeling.

Let us consider a classical problem of non-parametric regression:

$$y_i = f(x_i) + \varepsilon_i, \quad i = 1, \ldots, n,$$

where (ε_i) are independent, identically distributed variables, with zero mean and the variance σ^2.

If $f \in L^2(\mathbb{R})$ and ψ is the so-called mother wavelet, there are functions, called wavelets, defined by:

$$\psi_{jk}(x) = 2^{j/2}\psi(2^j x - k),$$

such as (ψ_{jk}) form an orthonormal basis for $f \in L^2(\mathbf{R})$ and, therefore, f can be written as:

$$f(x) = \sum_{k \in Z} c_k \phi_{0k}(x) + \sum_{j < J, j \in Z} d_{jk} \psi_{jk}(x),$$

where

$$c_k = \int_R f(x)\phi_{0k}(x)dx, \quad d_{jk} = \int_R f(x)\psi_{jk}(x)dx,$$

$$\phi_{j_0 k}(x) = 2^{j_0/2}\phi(2^{j_0}x - k),$$

ϕ is the father wavelets and J is a natural number giving the maximum resolution [20].

Practically, d_{jk} are firstly estimated and then the shrinkage is applied to the estimated values, $\tilde{d}_{jk}$, providing the new ones, $\hat{d}_{jk}$. The filtering process depends on a threshold, λ, as well as on the used filter-hard or soft.

In our studies [9, 11] we employed the hard threshold, which is defined by:

$$\delta_\lambda^H(d_{jk}) = d_{jk}I(|d_{jk}| > \lambda),$$

where I is the indicator function.

The main advantage of the wavelets procedure is the conservation of the coordinates d_{jk} that contain the significant information on the function f and the noise d_ε removal, following the decomposition rule:

$$d_{jk} = \hat{d}_{jk} + d_\varepsilon.$$

Stein's Unbiased Risk Estimate (SURE) and the universal threshold are the most used in applications [9, 11].

Finally, the function f is built as:

$$f(x) = \sum_{k=1}^{2^J-1} \hat{c}_k \phi_{0k}(x) + \sum_{j=0}^{J-1} \sum_{k=0}^{2^j-1} \hat{d}_{jk} \psi_{jk}(x),$$

where $\hat{c}_k$ is an estimation of c_k [19].

The alternative of the wavelets continuous transform (CWT) for discrete data is the discrete wavelets transform (DWT).

For example, DWT can be defined using the Daubechies wavelets, for a given the data sets $x_1, x_2, \ldots, x_n$, $n = 2^J$, following the steps [61]:

1. Consider the sets of coefficients:

$$\{h_0 = 1/\sqrt{2}, h_1 = 1/\sqrt{2}\}, \quad \{g_0 = 1/\sqrt{2}, g_1 = -1/\sqrt{2}\}.$$

2. Build the filters H, G and their duals, H*, G* such that:

$$\sum_k h_k = \sqrt{2}, \quad \sum_k g_k = 0, \quad H^*H + G^*G = Id,$$

   where Id is the unit matrix.
3. Build the continuous function f, defined by:

$$f = \sum_k f_k \phi_H(t - k),$$

   where

$$\phi_H(t) = \begin{cases} 1, & t \in [0, 1) \\ 0, & t \in \mathbf{R} - [0, 1) \end{cases}.$$

4. Use the CWT procedure for f built at the previous step.

For simulation using wavelets, **wavelets** and **wavethresh** packages from R can be used. For an extensive study on wavelets, readers can refers to [19, 20, 51, 52, 61].

References

1. Akaike, H.: Fitting autoregressive models for prediction. Ann. Inst. Stat. Math. **21**, 243–247 (1969)
2. Alder, H.L., Roessler, E.B.: Introduction to Probability and Statistics, 5th edn. W.H. Freeman and Company, San Francisco (1972)

3. Basak, D., Pal, S., Patranabis, D.C.: Support vector regression. Neural Inf. Process. Lett. Rev. **11**(10), 203–224 (2007)
4. Bărbulescu, A.: Time Series with Applications. Junimea, Iaşi (2002) (in Romanian)
5. Bărbulescu, A., Barbeş, L.: Mathematical models for inorganic pollutants in Constanţa area, Romania. Rev. Chim. **64**(7), 747–753 (2013)
6. Bărbulescu, A., Băutu, E.: Alternative models in precipitation analysis. Analele Ştiinţifice ale Universităţii Ovidius, Matematică **17**(3), 45–68 (2009)
7. Bărbulescu, A., Băutu, E.: Time series modeling using an adaptive gene expression programming. Int. J. Math. Models Meth. Appl. Sci. **3**(2), 85–93 (2009)
8. Bărbulescu, A., Băutu, E.: Mathematical models of climate evolution in Dobrudja. Theor. Appl. Climatol. **100**(1–2), 29–44 (2010)
9. Bărbulescu, A., Deguenon, J.: Models for trend of precipitation in Dobrudja. Environ. Eng. Manage. J. **13**(4), 873–880 (2014)
10. Bărbulescu, A., Maftei, C., Băutu, E.: Modeling the Hydro-Meteorological Time Series. Applications to Dobrudja Region. Lambert Academic Publishing, Germany (2010)
11. Bărbulescu, A., Petac, A.: Statistical assessment of precipitation evolution. Case Study. Autom. Comput. Appl. Math. **22**(1), 7–15 (2013)
12. Băutu, E., Bărbulescu, A.: Forecasting meteorological time series using soft computing methods: an empirical study. Appl. Math. Inf. Sci. **7**(4), 1297–1306 (2013)
13. Băutu, E., Băutu, A., Luchian, H.: AdaGEP—an adaptive gene expression programming algorithm. In: Proceedings of the Ninth international Symposium on Symbolic and Numeric Algorithms For Scientific Computing, pp. 403–406. SYNASC, IEEE Computer Society, Washington, DC, 26–29 Sept 2007
14. Bourbonais, J.: Cours et Exercices d'économétrie. Dunod, Paris (1993)
15. Box, G.P.E., Jenkins, G.M., Reinsel, G.C.: Time Series Analysis: Forecasting and Control, 4th edn. Wiley, Hoboken (2008)
16. Brockwell, P., Davies, R.: Introduction to Time Series Analysis and Forecasting. Springer, New York (2002)
17. Burg, J.P.: A new analysis technique for time series data. NATO Advanced Study Institute of Signal Processing. Enschede, Netherlands (1968)
18. Camps-Valls, G., Chalk, A.M., Serrano-López, A.J., Martín-Guerrero, J.D., Sonnhammer, E. L.L.: Profiled support vector machines for antisense oligonucleotide efficacy prediction. BMC Bioinform. **5**:135 (2004) http://www.biomedcentral.com/1471-2105/5/135
19. Chui, C.K.: An Introduction to Wavelets. Academic Press Inc., San Diego (1992)
20. Daubechies, I.: Ten lectures on Wavelets. SIAM, New Delhi (1992)
21. Deguenon, J., Bărbulescu, A.: Fractal characterization of rainfall in Benin. Int. J. Ecol. Econ. Stat. **30**(3), 46–55 (2013)
22. Dickey, D.A., Fuller, W.A.: Distribution of the estimators for autoregressive time series with a unit root. J. Am. Stat. Assoc. **74**(366), 427–431 (1979)
23. Ferreira, C.: Gene expression programming: a new adaptive algorithm for solving problems. Complex Syst. **13**(2), 87–129 (2001)
24. Ferreira, C.: Gene Expression Programming: Mathematical Modeling by an Artificial Intelligence. Springer, Berlin (2006)
25. Fuller, W.A.: Introduction to Statistical Time Series. Wiley, New York (1996)
26. Goldberg, D.E.: Genetic Algorithms in Search, Optimization, and Machine Learning. Addison-Wesley, Boston (1989)
27. Granger, C.W.J., Joyeux, R.: An introduction to long memory time series models and fractional differencing. J. Time Ser. Anal. **1**, 15–39 (1980)
28. Grossman, A., Morlet, J.: Decomposition of Hardy functions into square integrable wavelets of constant shape. SIAM J. Math. Anal. **1**, 723–736 (1984)
29. Haidu, I.: Time Series Analysis. Applications in Hydrology. HGA, Bucureşti (1997) (in Romanian)
30. Hannan, E.J., Risannen, J.: Recursive estimation of mixed autoregressive moving—average order. *Biometrika*, **69**(1), 81–94 (1981)

31. Hashimi, M.Z., Shamseldin, A.Y., Melville, B.W.: Statistical downscaling of watershed precipitation using Gene Expression Programming (GEP). Environ. Model. Softw. **26**(12), 1639–1646 (2011)
32. Haykin, S.: Neural networks and Learning Machines, 3rd edn. Pearson Education Inc., New York (2009)
33. Holland, J.H.: Adaptation in Natural and Artificial Systems: An Introductory Analysis with Applications to Biology, Control, and Artificial Intelligence. MIT Press, Cambridge (1992)
34. Hong, Y.-S.T., White, P.A., Scott, D.M.: Automatic rainfall recharge model induction by evolutionary computational intelligence. Water Resour. Res. **41**, W08422, 13 PP (2005)
35. Hsu, C.-W., Chang, C.-C., Lin, C.-J.: A practical guide to support vector classification. Technical report, Department of Computer Science and Information Engineering, National Taiwan University, Taipei (2010) http://www.csie.ntu.edu.tw/~cjlin/papers/guide/guide.pdf
36. https://cran.r-project.org/web/packages/forecast/forecast.pdf
37. https://cran.r-project.org/web/packages/fUnitRoots/fUnitRoots.pdf
38. https://cran.r-project.org/web/packages/urca/urca.pdf
39. http://faculty.washington.edu/ezivot/econ584/notes/unitroot.pdf
40. https://stat.ethz.ch/R-manual/R-devel/library/stats/html/arima.sim.html
41. http://www.support-vector-machines.org/SVM_soft.html
42. Hung, N.Q., Babel, M.S., Weesakul, S., Tripathi, N.K.: An artificial neural network model for rainfall forecasting in Bangkok, Thailand. Hydrol. Earth Syst. Sci. **13**, 1413–1425 (2009)
43. Hyndman, R., Athanasopoulos, G.: Forecasting: Principles and Practice. OTexts: Melbourne (2014) https://www.otexts.org/fpp/8/7
44. Hyndman, R.J., Khandakar, Y.: Automatic time series forecasting: the forecast package for R. J. Stat. Softw. **26**(3) (2008)
45. Kaneko, H., Funatsu, K.: Application of online support vector regression for soft sensors. AIChE J. **60**, 600–612 (2014)
46. Koza, J.R.: Genetic programming: on the programming of computers by means of natural selection. MIT Press Cambridge, Massachusetts (1992)
47. Kwiatkowski, D., Phillips, P.C.B., Schmidt, P., Shin, Y.: Testing the null hypothesis of stationarity against the alternative of a unit root. J. Econometrics **54**, 159–178 (1992)
48. Langdon, W.B., Poli, R.: Foundations of Genetic Programming. Springer, Berlin (2002)
49. McCullogh, W., Pitts, W.: A logical calculus of the ideas immanent in nervous activity. Bull. Math. Biophys. **5**, 115–133 (1943)
50. MacKinnon, J.G.: Numerical distribution functions for unit root and cointegration tests. J. Appl. Econometrics **11**, 601–618 (1996)
51. Nason, G.P.: Choice of the threshold parameter in wavelet function estimation. In: Antoniadis, A., Oppenheimer, G. (eds.) Wavelets and Statistics, Lecture Notes in Statistics, vol. 103, pp. 261–280. Springer, Berlin (1995)
52. Nason, G.P.: Wavelets Methods in Statistics with R. Springer, Berlin (2008)
53. Pfaff, B.: Analysis of Integrated and Cointegrated Time Series with R, 2nd edn. Springer, Berlin (2008)
54. Phillips, P.C.B., Perron, P.: Testing for a unit root in time series regression. Biometrika **75**(2), 335–346 (1998)
55. Sapankevych, N., Sankar, R.: Time series prediction using support vector machines: a survey. IEEE Comput. Intell. Mag. **4**(2), 24–38 (2009)
56. Schmidt, P., Phillip, P.C.B.: LM tests for a unit root in the presence of deterministic trends. Oxford Bull. Econ. Stat. **54**(3), 257–287 (1992)
57. Schwarz, G.E.: Estimating the dimension of a model. Ann. Stat. **6**(2), 461–464 (1978)
58. Smola A.J., Schölkopf B.: A tutorial on support vector regression. Statistics and Computing, **14**, 199–222 (2004)
59. Specht, D.F.: General regression neural network. IEEE Trans. Neural Netw. **2**(6), 568–576 (1991)
60. Specht, D.F.: Enhancements to probabilistic neural networks. Int. Jt. Conf. Neural Netw. **I**, 761–768 (1992)

61. Stark, H.-G.: Wavelets and Signal Processing. Springer, Berlin (2005)
62. Tay, F.E.H., Cao, L.: Application of support vector machines in financial time series forecasting. Omega: Int. J. Manage. Sci. **29**(4), 309–317 (2001)
63. Vapnik, V.: The Nature of Statistical Learning Theory. Springer, New York (2000)

Chapter 3
Models for Precipitation Series

The aspects investigated in this chapter can be resumed as follows:

- Building ARMA models for precipitation series in Dobrogea region;
- Generation of annual precipitation series based on ARMA models;
- Generation of monthly precipitation series using MMF and AAS.
- Building decomposition models for two series.
- Comparisons of performances of different algorithms from artificial intelligence for forecasting precipitation series. It will be emphasized that the Adaptive Gene Expression Programming algorithm performs better than the other procedures on the studied cases.
- Comparison of decomposition and wavelets models for determination the trend of a series; case studies.

1 ARMA Models for Precipitation Series and Generation of Precipitation Fields

1.1 ARMA Models for Precipitation Series and Generation of Annual Precipitation Fields

After '70s when the Box-Jenkins methodology has been introduced, ARIMA and other models derived from it have been extensively studied due to their applications in different branches of sciences, as meteorology, climatology, hydrology, finance etc., for modeling and forecasting.

Researchers proved that the Box-Jenkins methodology can be successfully used to fit data that do not present very high variability and jumps. Our studies, conducted in the direction of finding best models for temperature and precipitation series in Dobrogea region, confirmed these results.

© Springer International Publishing Switzerland 2016
A. Bărbulescu, *Studies on Time Series Applications in Environmental Sciences*,
Intelligent Systems Reference Library 103, DOI 10.1007/978-3-319-30436-6_3

ARIMA models for the evolution of temperature and precipitation in the south-eastern part of Romania are presented in [1, 5, 6, 9, 10].

Even if ARIMA models do not always provide the expected results, they can be used, combined with artificial intelligence methods, in hybrid techniques, that successfully solve the problem at hand [2].

Generating annual precipitation series has a reduced number of direct applications, but it is useful for two purposes. Firstly, it facilitates the understanding of the stochastic nature of annual precipitation and the implications of its absence or of the existence of long periods of heavy rainfall. This knowledge is the base of the water resources management in dry periods. Secondly, a good annual precipitation model, that preserves the characteristics of the precipitation series at different time scales, could be used for the disaggregation of annual precipitation in monthly data which become the input of different disaggregation schemes [38].

To generate annual rainfall, Srikanthan and McMahon [36] used an AR(1) model. This model has been criticized for its inability to explicitly describe the dry years and drought in the observed data. As a result, Thyer and Kuczera [44] developed a Bayesian inference model to generate annual precipitation for Sydney, Australia. It assumes that the climate consists of two states, each one having a normal distribution of rainfall. The transition from one state to the other is described by transition probabilities [33]. The models' parameters are estimated using MCMC, yielding their a posteriori probability distribution.

After applying both methods on 44 precipitation series in Australia, it resulted that the new proposed method gave better results in all situations.

After taking into account the parameters' uncertainty both algorithms had similar performances [35].

Given a sample of data $x_t, t = \overline{1, n}$, let us denote by:

$$r = \frac{1}{(n-1)s^2} \sum_{t=1}^{n-1} (x_{t+1} - \bar{x})(x_t - \bar{x})$$

the lag one autocorrelation coefficient, $\bar{x}$, the average of the sample, s, its standard deviation and g, the skewness:

$$g = \frac{n}{(n-1)(n-2)s^3} \sum_{t=1}^{n-1} (x_{t+1} - \bar{x})(x_t - \bar{x}).$$

To generate the annual precipitation at a site, the following AR(1) model has been proposed and used [35, 38]:

$$X_t = rX_{t-1} + \sqrt{1 - r^2}\eta_t, \tag{1}$$

where X_t is the standardized precipitation in the period t and η_t is a random variable with a standard normal distribution.

In this model, the volume of annual precipitation is estimated by:

$$x_t = \bar{x} + sX_t. \tag{2}$$

In the case of skewed data (with the skewness coefficient greater than 0.5), their asymmetry can be modeled by the Wilson–Hilferty transformation:

$$\varepsilon_t = \frac{2}{g_\varepsilon}\left[\left(1 + \frac{1}{6}g_\varepsilon\eta_t - \frac{1}{36}g_\varepsilon^2\right)^3\right] - 1,$$

where g_ε is the skewness of ε_t, related to that of the annual data (g) by:

$$g_\varepsilon = \frac{1 - r^3}{(1 - r^2)^{3/2}}g.$$

Another model for generation annual Gaussian precipitation data is:

$$X_t = \alpha + \beta X_{t-1} + e_t, \tag{3}$$

where the unknown parameters are α, β and σ_ε^2 (the variance of e_t).

The parameters α, β are estimated by the least squares method, the annual data are generated by sampling σ_ε^2 from a chi-square distribution with $(n-3)$ degrees of freedom, and $\theta = (\alpha, \beta)$, from a Gaussian distribution whose parameters were previously estimated [35].

For taking into account the parameters' uncertainty, another AR(1) model can be used for generation Gaussian precipitation data. It has the equation:

$$z_t = \mu + \varphi_1(z_{t-1} - \mu) + \varepsilon_t, \tag{4}$$

where (z_t) has the mean μ, and ε_t is a normal random variable, with the zero mean and unit variance.

If the data are not Gaussian, they are normalized using a Box–Cox transformation:

$$z_t = \begin{cases} \frac{x_t^\lambda - 1}{\lambda}, & \lambda \neq 0 \\ \log x_t, & \lambda = 0 \end{cases}, \tag{5}$$

where λ is the transformation parameter, that must be determined. Including it, the $\theta^T = (\mu, \sigma_\varepsilon, \varphi_1, \lambda)$ is the vector that must be determined in AR(1) model. To simulate the precipitation data, the Metropolis algorithm can be utilized, as in [34].

The utilization of another kind of models is recommended in [35, 38] for the situation when the results of the previous models are not satisfactory. Also, it is recommended to repeat the experiment of annual precipitation field generation for minim 100 times, followed by determination of the mean and the results' comparison.

1.1.1   Generation of Annual Precipitation Series for the Main Stations in Dobrogea

In this section, we present the results of the application of some methods already presented for generating the annual precipitation fields from different sites in Dobrogea.

Firstly, the values of the basic statistics for data series have been determined (Table 1) and different tests have been performed (at the significance level of 0.05).

The Kolmogorov–Smirnov, Shapiro–Wilk, Jarque–Bera and Anderson–Darling tests led to the rejection of normality hypothesis for Corugea and Harsova series [7]. The Jarque–Bera test also rejected the normality hypothesis for Mangalia (Table 2). For the series LogCor and LogHar, obtained by taking logarithms of the data series Corugea and Harsova, the normality hypothesis could not be rejected.

The value of the skewness coefficient (Table 1) emphasized the existence of a significant asymmetry for the majority of series.

Table 1   Values of basic statistics for annual series

	Min	Max	Mean	Median	Standard deviation	Coef. of variation	Skewness	Kurtosis
Adamclisi	297	732	485	479	118	0.244	0.297	2.1108
Cernavoda	255	844	488	475	128	0.263	0.802	3.9666
Constanta	227	675	423	411	110	0.257	0.487	2.8357
Corugea	271	815	435	410	115	0.266	1.38	5.0215
Hârsova	226	803	409	375	137	0.336	1.14	3.7570
Jurilovca	203	687	378	376	110	0.290	0.535	3.1891
Mangalia	250	758	428	409	107	0.250	0.886	4.0896
Medgidia	223	714	450	450	111	0.246	0.384	2.9774
Sulina	110	487	262	252	78.2	0.299	0.430	3.2651
Tulcea	274	732	462	110	443	0.237	0.592	2.7716

Table 2   p-values of the test-statistics in the Jarque–Bera and Anderson–Darling tests

	Jarque–Bera	Anderson–Darling
Adamclisi	0.3946	0.2445
Cernavoda	0.0587	0.1667
Constanta	0.4610	0.1177
Corugea	7.058e−05	0.0038
Harsova	0.0097	0.0024
Jurilovca	0.3915	0.5333
Mangalia	0.0301	0.1735
Medgidia	0.6267	0.8021
Sulina	0.5235	0.8659
Tulcea	0.3155	0.0682

The skewness and kurtosis have been computed using the R package **moments** [16], writing the following code:

```
library(moments)
skewess(x)# x is the series to which we apply the test
kurtosis(x)
```

The analysis of the autocorrelation function and partial autocorrelation functions of data series proved that the only series that presents a first order autocorrelation and partial autocorrelation is Sulina (Fig. 1).

Taking into account only the normality and the autocorrelation tests, it results that Adamclisi, Constanta, Cernavoda, Log(Cor), Log(Har), Jurilovca, Mangalia, Medgidia, Tulcea are not autocorrelated and are Gaussian, so independente.

Applying the Box–Jenkins methodology and the Akaike selection criterion, the models presented in Table 3 have been determined for these series. They are: a Gaussian white noise (for Adamclisi), an AR(2) model (for Mangalia), and AR(1) models for the rest of the series, but the residual variances are very high. Also, the skewness coefficient is high, so generation precipitation fields based on these models' can not give satisfactory results.

Fig. 1 ACF şi PACF of Sulina series

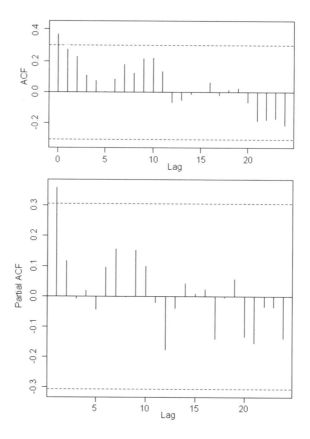

Table 3 Models for annual precipitation series at the main stations

Series	Tests' results		ARMA models	
	ACF–PACF	AIC	Model	AIC
Adamclisi	Gaussian noise Variance = 13,998.12	133.7	Gaussian noise Variance(Z_t) = 13,998.12	133.7
Cernavoda	Gaussian noise Variance = 16,477.28	112.2	$X_t = 0.9397\,X_{t-1} + Z_t$ Variance(Z_t) = 24,529.9	537.2
Constanta	Gaussian noise Variance = 12,012.4	616.9	$X_t = 0.9583\,X_{t-1} + Z_t$ Varianţa Z_t = 18,668	526.4
Corugea	Non-Gaussian, uncorrelated Variance = 13,340.17	619.3	$X_t = 0.9668\,X_{t-1} + Z_t$ Variance(Z_t) = 17,415.1	523.7
Log(Cor)	Gaussian noise Variance = 0.01	197.6	$X_t = 0.9716\,X_{t-1} + Z_t$ Variance(Z_t) = 0.032233	−17.3
Harsova	Gaussian noise Variance = 18,858.99	615.8	$X_t = 0.9319\,X_{t-1} + Z_t$ Variance(Z_t) = 29,384.1	544.5
Log(Har)	Gaussian noise Variance = 0.02	196.6	$X_t = 0.9706\,X_{t-1} + Z_t$ Variance(Z_t) = 0.044	−3.98
Jurilovca	Gaussian noise Variance = 12,024.92	608.4	$X_t = 0.9507\,X_{t-1} + Z_t$ Variance(Z_t) = 16,032	520.0
Mangalia	Gaussian noise Variance = 11,418.62	617.7	$X_t = 0.5225\,X_{t-1} + 0.418\,X_{t-2} + Z_t$ Variance(Z_t) = 16,562.4	523.3
Medgidia	Gaussian noise Variance = 12,231.08	621.7	$X_t = 0.9646\,X_{t-1} + Z_t$ Variance(Z_t) = 17,717.3	524.4
Sulina	Gaussian, correlated Variance = 6117.6	578.4	$X_t = 0.9484\,X_{t-1} + Z_t$ Variance(Z_t) = 7624.31	489.5
Tulcea	Gaussian noise Variance = 12,013.05	623.7	$X_t = 0.9654\,X_{t-1} + Z_t$ Variance(Z_t) = 18,858.8	527.0

Therefore, we looked for models described by (4), for the normalized data obtained using a transformation (5), with λ given in Table 4.

The models of the new series are presented in Table 5. Based on these models, 100 samples have been generated, each of them containing the same number of data as the historical series. Then the average of the parameters in these models and different percentile have been computed using SCL software [18].

The data generated should have the same characteristics as the recorded ones.

Each sample generated based on a model is different and its characteristics differ from those of the record data, but the average of each characteristic of all the samples should be equal to that of the record data.

The comparison between the registered and generated data is done function of an established tolerance (Table 6) [18].

Figures 2, 3 and 4 present the registered and the simulated data, together with their percentiles at 2.5, 50 and 97.5 % for the annual series Cernavoda, Constanta and Medgidia.

Table 4 The values of the parameter λ in the Box–Cox transformation for the annual precipitation series

	Adamclisi	Cernavoda	Constanta	Corugea	Harsova	Jurilovca	Mangalia	Medgidia	Sulina	Tulcea
λ	0.175	0.130	0.252	−0.895	−0.506	0.252	−0.126	0.443	0.513	−0.061

Table 5 Models for the transformed series

Series	Model	Residual variance	AIC
Adamclisi	$X_t = 0.142\ X_{t-1} + Z_t$	0.5067	92.669
Cernavoda	$X_t = 0.169\ X_{t-1} + Z_t$	0.3189	73.631
Constanta	$X_t = 0.146\ X_{t-1} + Z_t$	1.3515	133.185
Corugea	$X_t = 0.115\ X_{t-1} + Z_t$	0.00001	−442.827
Harsova	$X_t = 0.127\ X_{t-1} + Z_t$	0.0002	−225.403
Jurilovca	$X_t = 0.307\ X_{t-1} + Z_t$	1.4915	137.033
Mangalia	$X_t = -0.141\ X_{t-1} + Z_t$	0.012	−59.284
Medgidia	$X_t = 0.187\ X_{t-1} + Z_t$	12.879	224.988
Sulina	$X_t = 0.388\ X_{t-1} + Z_t$	22.627	248.675
Tulcea	$X_t = 0.1637\ X_{t-1} + Z_t$	0.0248	−30.8730

Table 6 Tolerance of statistics in annual precipitation generation

Statistics	Tolerance
Mean	5 %
Standard deviation	5 %
Skewness	0.5 %
Autocorrelation coefficient of first order	0.15 %
Max, min, range	10 %
Sum of low precipitation on 2 (3, 5, 7, 10) years	10 %

Fig. 2 Historical and simulated data for Cernavoda annual series

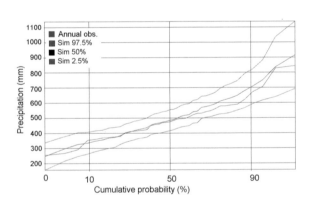

Tables 7, 8, 9 and 10 contain the results of simulation the precipitation annual precipitation from some main stations.

The standard deviation is not estimated in the established tolerance limits, especially due to the high variance after the 95 % percentile. For Sulina, the mean and the 5 years—sum low precipitation are also outside the tolerance limits and for Harsova and Corugea, one, respectively two statistics values. However, the results of generation the annual precipitation fields at a single site are satisfactory.

Fig. 3 Historical and
simulated data for Constanta
annual series

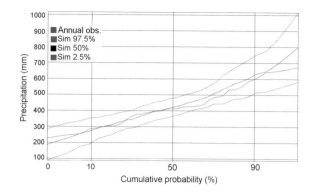

Fig. 4 Historical and
simulated data for Medgidia
annual series

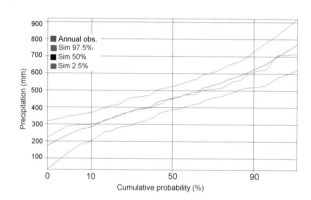

Table 7 Results of simulation for Corugea annual series

	Historical	Mean	2.50 %	50 %	97.50 %	Yes/no
Mean	434.67	454.39	396.432	447.14	534.45	Yes
Standard deviation	115.5	158.32	77.733	127.17	430.75	No
Skewness	1.384	1.582	0.093	1.257	4.243	Yes
1st order autocorrel. coef.	0.208	0.085	−0.329	0.071	0.544	Yes
Max	1.874	2.253	1.428	1.865	5.02	No
Min	0.624	0.592	0.42	0.59	0.72	Yes
2 year—sum low prec	1.454	1.342	1.022	1.367	1.59	Yes
3 year—sum low prec	2.395	2.159	1.704	2.164	2.55	Yes
5 year—sum low prec	4.031	3.889	3.113	3.971	4.434	Yes
7 year—sum low prec	5.856	5.706	4.611	5.884	6.399	Yes
10 year—sum low prec	8.637	8.51	6.917	8.71	9.373	Yes

Table 8 Results of simulation for Harsova annual series

	Historical	Mean	2.50 %	50 %	97.50 %	Yes/no
Mean	408.82	439.47	366.649	427.66	577.55	No
Standard deviation	137.32	159.42	95.1	153.09	551.19	No
Skewness	1.143	1.588	0.221	1.197	4.033	Yes
1st order autocorrel. coef.	0.134	0.095	−0.223	0.096	0.469	Yes
Max	1.963	2.549	1.611	2.097	5.981	No
Min	0.552	0.493	0.31	0.499	0.645	No
2 year—sum low prec	1.181	1.158	0.771	1.166	1.489	Yes
3 year—sum low prec	2.031	1.899	1.311	1.926	2.305	Yes
5 year—sum low prec	3.523	3.535	2.463	3.588	4.258	Yes
7 year—sum low prec	5.271	5.275	3.737	5.361	6.198	Yes
10 year—sum low prec	7.906	8.034	5.736	8.169	9.13	Yes

Table 9 Results of simulation for Mangalia annual series

	Historical	Mean	2.50 %	50 %	97.50 %	Yes/no
Mean	425.30	435.36	391.45	433.314	513.493	Yes
Standard deviation	103.57	117.65	78.84	109.699	188.44	No
Skewness	0.908	0.758	−0.328	0.695	2.121	Yes
1st order autocorrel. coef.	−0.152	−0.165	−0.515	−0.211	0.282	Yes
Max	1.781	1.754	1.384	1.64	2.486	Yes
Min	0.587	0.566	0.395	0.568	0.708	Yes
2 year—sum low prec	1.51	1.381	1.014	1.393	1.621	Yes
3 year—sum low prec	2.264	2.269	1.786	2.297	2.579	Yes
5 year—sum low prec	4.059	4.095	3.488	4.143	4.535	Yes
7 year—sum low prec	5.864	5.995	5.202	6.042	6.455	Yes
10 year—sum low prec	8.4	8.897	7.853	8.972	9.519	Yes

Table 10 Results of simulation for Sulina annual series

	Historical	Mean	2.50 %	50 %	97.50 %	Yes/no
Mean	261.634	274.791	217.66	270.35	360.58	No
Standard deviation	78.215	85.421	52.013	79.38	150.05	No
Skewness	0.43	0.382	−0.595	0.353	1.756	Yes
1st order autocorrel. coef.	0.359	0.339	0.019	0.313	0.696	Yes
Max	1.861	1.768	1.425	1.717	2.404	Yes
Min	0.419	0.426	0.095	0.455	0.638	Yes
2 year—sum low prec	1.043	1.046	0.475	1.093	1.382	Yes
3 year—sum low prec	1.759	1.783	0.949	1.831	2.331	Yes
5 year—sum low prec	3.064	3.411	2.084	3.518	4.235	No
7 year—sum low prec	4.735	5.171	3.66	5.253	6.062	Yes
10 year—sum low prec	8.095	7.963	5.91	8.102	9.097	Yes

1.1.2   Models for Annual Precipitation Series from the Secondary Stations in Dobrogea

After the preliminary statistical analysis, the following results have been obtained (Table 11).

- The normality hypothesis could not be rejected for 21 series and has been rejected for the rest. However, after different transformations (Table 11, column 2), the normality hypothesis has been rejected only for three series.
- The homoskedasticity tests could not reject the null hypothesis for all series, but Albesti and Saraiu.
- The correlograms emphasized the existence of different order of autocorrelation only for the series: Altan-Tepe, Amzacea, Casimcea, Pantelimon, Pietreni—1st order, Albeşti—2nd order, Casian—3rd order.
- The break tests revealed the breakpoint existence for the majority of series, especially in 2003.

Since the skewness coefficient was greater than 0.2, Box-Cox transformations have been applied to correct it before the modeling stage. The values of the parameter λ in the transformations are presented in the third column of Table 11. Finally, the ARMA models have been determined for the transformed series. Their equations are given in the last column of Table 11.

The precipitation fields generated for the secondary series based on the AR(1) models describes well the precipitation evolution at these sites.

Table 11 Results of statistical tests and models for secondary series

Series	Normality	λ	Model
Agigea	No	−0.191	$X_t = Z_t - 0.1641\ Z_{t-1}$
Albesti	Yes (after radical extraction)	−0.018	$X_t = 0.3864\ X_{t-2} + Z_t$
Altân Tepe	Yes	−0.065	$X_t = Z_t + 0.4823\ Z_{t-1}$
Amzacea	No	0.240	$X_t = 0.3087\ X_{t-1} + Z_t$
Baia	Yes	0.602	$X_t = Z_t - 2.165\ Z_{t-4}$
Băltăgeşti	Yes	0.366	$X_t = Z_t + 0.5549\ Z_{t-2} + 0.2619\ Z_{t-3}$
Biruinţa	Yes (after radical extraction)	−0.387	$X_t = Z_t - 0.277\ Z_{t-5}$
Casian	Yes	−0.315	$X_t = Z_t - 0.4320\ Z_{t-3}$
Casimcea	Yes (after taking log)	0.098	$X_t = Z_t + 0.5402\ Z_{t-1}$
Ceamurlia	Yes	0.758	$X_t = 0.2663\ X_{t-1} + Z_t$
Cerna	Yes	0.009	$X_t = 0.4641\ X_{t-1} + Z_t - 0.8295\ Z_{t-1}$
Cheia	Yes (after radical extraction)	−0.154	$X_t = 0.2691\ X_{t-1} + Z_t$
Cobadin	Yes	0.523	$X_t = 0.2234\ X_{t-1} + Z_t$
Corbu	Yes (after radical extraction)	−0.196	$X_t = -0.327\ X_{t-1} + Z_t$
Crucea	Yes	0.105	$X_t = 0.1130\ X_{t-1} + Z_t$
Cuza Vodă	Yes	0.119	$X_t = Z_t + 0.2289\ Z_{t-1}$
Dăieni	Yes (after taking log)	−0.370	$X_t = -0.4205\ X_{t-2} + Z_t + 0.7567\ Z_{t-2}$
Dobromir	Yes	0.440	$X_t = Z_t + 0.4830\ Z_{t-2}$
Dorobanţu	Yes (after taking log)	−0.594	$X_t = Z_t + 0.3557\ Z_{t-1}$
Greci	Yes	0.617	$X_t = Z_t - 0.3197\ Z_{t-2}$
Hamcearca	Yes (after radical extraction)	−0.393	$X_t = 0.3999\ X_{t-3} + Z_t$
Independenţa	Yes	0.182	$X_t = 0.3018\ X_{t-1} + Z_t$
Lipniţa	Yes (after radical extraction)	−0.052	$X_t = 0.3949\ X_{t-2} + Z_t$
Lumina	No	0.650	$X_t = 0.1808\ X_{t-1} + Z_t - 0.2212\ Z_{t-1}$
Mihai Viteazu	Yes	0.330	$X_t = 0.2449\ X_{t-1} + Z_t$
Negru Vodă	Yes (after radical extraction)	0.258	$X_t = 0.1162\ X_{t-1} + Z_t$
Negureni	Yes	−0.382	$X_t = Z_t - 0.1778\ Z_{t-1}$
Niculiţel	Yes	−0.195	$X_t = 1.33\ X_{t-1} - 0.6563\ X_{t-2} + Z_t - 1.796\ Z_{t-1} + 0.5632\ Z_{t-2}$
Nuntaşi	Yes	1.060	$X_t = 0.1005\ X_{t-1} + Z_t$
Pantelimon	Yes (after taking log)	−0.591	$X_t = 0.4401\ X_{t-1} + 0.2934\ X_{t-2} + Z_t$

(continued)

Table 11   (continued)

Series	Normality	λ	Model
Peceneaga	Yes (after taking log)	−0.880	$X_t = Z_t - 0.4974\ Z_{t-2}$
Pecineaga	Yes (after radical extraction)	−0.597	$X_t = -0.02760\ X_{t-1} + Z_t$
Peștera	Yes (after taking log)	−0.438	$X_t = -0.4425\ X_{t-3} + Z_t$
Pietreni	Yes	0.470	$X_t = 0.3322\ X_{t-1} + 0.3121\ X_{t-2} + Z_t$
Poșta	Yes (after radical extraction)	−0.416	$X_t = 0.4242\ X_{t-1} + Z_t$
Săcele	Yes	−0.619	$X_t = 0.2884\ X_{t-1} + Z_t$
Saraiu	Yes (after a Box-Cox transf.)	0.145	$X_t = Z_t + 0.08966\ Z_{t-1}$
Satu Nou	Yes	0.923	$X_t = Z_t + 0.2999\ Z_{t-1}$
Siliștea	Yes	0.337	$X_t = Z_t - 0.7158\ Z_{t-5}$
Topolog	Yes (after taking log)	−0.895	$X_t = Z_t + 0.08700\ Z_{t-1}$
Zebil	Yes	0.126	$X_t = Z_t + 0.3231\ Z_{t-1}$

1.2   Generation of Monthly Precipitation Series

1.2.1   Theoretical Considerations

Monthly precipitation series can be generated by using: autoregressive processes (AR), the Thomas-Fiering model (TF), the fragments' method (F) and its modified version (MF), the two-tier model (TT), wavelets etc. [45].

The fragments method, introduced by Srikanthan and McMahon [37], relies on the disaggregation of the annual precipitation generated by an AR(1) model. Porter and Pink [26] proposed the obtaining of monthly fragments from a flow sequence generated monthly, but the method doesn't preserve the correlation between the first month and the last one of the previous year.

Maheepala and Perera [23] modified this procedure such that the monthly correlation along the consecutive years is preserved, but the correlation between the first month and the last one of the previous year is still not kept [30]. This inconvenience can be removed considering that the beginning of the hydrological year is the month with the minimum serial correlation. Also, it has been seen that the annual pattern is repeated if only a reduced number of annual data is available [36].

This method has been modified by introducing synthetic fragments [39] and has been used for generation of monthly precipitation at ten sites from Australia. It seems that it doesn't give satisfactory results if there are months without precipitation. The modified method of fragments (MMF) is described in the following [39].

- The monthly data are generated by the Thomas–Fiering model;
- The annual data are generated;
- The record monthly data are standardized, by dividing the monthly value from a year by the annual precipitation. Therefore, the sum of standardized precipitation from each year is 1. So, the number of fragments of monthly precipitation is equal to the number of years.
- The annual data generated at the second step are disaggregated using the fragments produced at the first stage. The monthly fragments of a year, k, are selected taking into account the degree of closeness of the annual precipitation generated by the monthly precipitation from the last month of the previous year to the data already disaggregated corresponding to the historical values [23]. This process is realized by the selection of the monthly fragments of the ith year in the monthly series generated, that minimizes the sum

$$\alpha_1 = \left(\frac{x'_k - x_i}{s_x}\right)^2 + \left(\frac{y'_{k-1} - y_{i-1}}{s_y}\right)^2,$$

where:

x'_k the monthly precipitation generated for the year k,
x_i the historical annual precipitation from the year i,
s_x the standard deviation of the annual precipitation,
y'_{k-1} the monthly disaggregated precipitation from the last month of the year $k - 1$,
y_{i-1} the historical monthly rainfall from the last month of the year $i - 1$,
s_y the standard deviation of the annual precipitation of the last month of the year.

The generated annual precipitation is disaggregated by multiplying the general precipitation with each of the 12 fragments, for obtaining 12 series of monthly precipitation.

The equation of the Thomas–Fiering model [42] is:

$$x_{i+1} = \bar{x}_{j+1} + b_j(x_i - \bar{x}_j) + t_i s_{j+1}\sqrt{1 - r_j^2},$$

where:

x_i, x_{i+1} are the values of the precipitation generated in the months i, and $i + 1$,
$\bar{x}_j, \bar{x}_{j+1}$ are the averages of precipitation from the month j and $j + 1$,
b_j is the regression coefficient,
t_i is a random variable normally distributed, with zero mean and unit standard deviation,
s_{j+1} is the standard deviation of precipitation in the month $j + 1$,
r_j is the correlation coefficient between the rainfall in the month j and $j + 1$.

Since the precipitation series are not generally Gausssian, the use of a random variable t_i with a gamma distribution with three parameters has been proposed. In Two-Tier model, the annual data are generated by using wavelets, and the monthly ones, by the Thomas–Fiering model, with a gamma variable. The monthly data are then adjusted such that their sum coincides with the annual ones, by the formula:

$$x'_{i,j} = \frac{x_{i,j}}{\sum_{j=1}^{12} x_{i,j}},$$

where:

x' is the adjusted generated monthly precipitation,
x is the generated monthly precipitation,
i, j are the annual and monthly indices.

Thompson [43] proposed a model based on Poisson process of rainfall apparition, in which the precipitation quantity is exponentially distributed.

A nonparametric model (NP2) that preserves the short term (monthly) and inter-yearly correlations is presented in [39]. It uses a Gaussian kernel for estimating the probability density.

Disaggregation scheme are also utilized to generate monthly rainfall at many sites simultaneously [20, 24, 40], some of them being nonparametric [27, 41].

To present these methods, we emphasize that in the stochastic disaggregation, a value recorded at a superior level is decomposed in many values at an inferior level, preserving the statistical characteristics at both levels.

If d is the number of seasons (12, for monthly data and 4, for quarterly data), denote by $Y = (Y_1, \ldots, Y_d)^T$ the variables at the lower and by X those at the upper level (annual). Then, the sum of Y_i must be X.

Koutsoyiannis and Manetas [20] developed a procedure (APP) that involves two different models, one for the lower level and the other one, for the upper level, independently developed, as follows:

(1) An ARMA model is generated for the annual data;
(2) A periodic ARMA model is generated for the monthly data, independent from that generated at step (1).
   Let denote by $\hat{Y} = (\hat{Y}_1, \ldots, \hat{Y}_d)^T$ the estimated vector;
(3) The sum of the components of $\hat{Y}$, denoted by $\hat{X}$, is compared with the value generated at step (1), X. Denoting by σ_X the standard deviation of X, $\varepsilon \in (0.1, 1)$—a tolerance and $\Delta = |X - \hat{X}|/\sigma_X$, verify if Δ is greater than ε. In affirmative case, $\hat{Y}$ is generated again. If not, the values of $\hat{Y}$ are adjusted to satisfy the additivity condition, using a linear, proportional or power scheme.
(4) Repeat the steps (1)–(3) until finishing the disaggregation.

The linear scheme preserves the average and the variance–covariance matrix of the variables at the inferior level, but it can generate negative values. Also, the statistics of higher orders are not always preserved.

The proportional scheme is recommended if the variables at the inferior level are independent [22], with marginal gamma distributions and the same scale parameter.

The disaggregation nonparametric procedure proposed in [41] can not be easily implemented because it uses a multivariate kernel. Therefore, a new approach has been proposed in [27]. The coordinates of the vector Z are generated by the rotation of Y in a new space. It is done utilizing the rotation matrix R, obtained by the Gramm–Schmidt orthogonalization process, such that $Z = RY$.

The new algorithm (NPDK) has the following steps:

(1) Estimation of the matrix R from the historical data, resulting in $z_i = (z_{i,1}, \ldots, z_{i,d})$, $i = \overline{1, n}$, where n is the number of years;
(2) Generation of the data at the upper level using a model specified by the user and the computation of $Z_d = X/\sqrt{d}$;
(3) Choosing the closest k neighbours (k-NN) by the estimation of the distances between Z_d and $z_{i,d}$.
(4) Building the vector Z, whose first $d - 1$ elements are those from the step (1) and the last one is Z_d;
(5) Rotate Z in the original space;
(6) Repeat the steps (2)–(5) until the generation of all data.

Lee et al. [22] proposed a disaggregation procedure that uses parts of the two algorithms presented. KNNR is employed for finding the values from the lower level whose sum is close to the values previously generated at the upper level. Then, the adjusting procedure is applied, to fulfill the additivity condition.

The original idea of this approach consists in the inclusion of the value from the last season of the previous year in the choice of the sequence at the lower level and the use of a genetic algorithm for detecting the variable at that level.

A competitive monthly multisite model must be capable of reproducing the relevant properties of the observed precipitation, as the spatial and temporal dependence, the marginal distribution functions (at each station) at monthly and annual scale. It also must preserve the properties of the extreme events (intensity, severity, duration, the distance between the events), etc. [29].

Generation of monthly precipitation at many sites can be done by the disaggregation of the annual precipitation generated using the fragments methods [37], the method of synthetic fragments [26] or the modified method of synthetic fragments [23], by minimizing the sum:

$$\sum_{j=1}^{N} \left(\frac{x_k^j - x_i^j}{s_x^j} \right)^2 + \sum_{j=1}^{N} \left(\frac{y_{k-1}^j - y_{i-1}^j}{s_y^j} \right)^2,$$

where N is the number of stations, and j is the site index.

The extended model of Mejia and Rousselle [24], used for the disaggregation of the annual flows in monthly flows, has the equation:

$$Y = AX + B\varepsilon + CZ,$$

where Z is a column vector that contains the monthly values from the previous years and A, B, C are matrices of coefficients.

Another model [21] has the equation for the ith month:

$$Y_{t,i} = A_i X_t + B_t \varepsilon + C_i Y_{t,i-1}.$$

The biggest problem of these models remains the high number of parameters to be estimated when working with a big number of series. Therefore, the scientists proposed simplifications of the problem at hand. The variations of the model's parameters from year to year must be considered in all cases [33].

A complex approach of modeling the multisite precipitation appears in [29]. The algorithm contains different modules that can be combined function of the record.

The first module contains the deseasonalisation scheme. The deseasonalized residuals are obtained by

$$DR_i(t) = (LR_i(t) - m_{LR,i})/s_{LR,i}, \quad i, t \in \overline{1, n},$$

where $R(t)$, $t = \overline{1, n}$ represents the historical data, $m_{LR,i}$ is the average of each month, $LR(t) = \log R(t)$, and $s_{LR,i}$ is the monthly standard deviation.

The second module performs the bootstrap resampling. If the deseasonalized residuals are approximately independent in time but spatially correlated, the simultaneous resampling is used for simulation.

The third module contains the *GAMLSS* modeling. If the deseasonalized residuals are correlated, then the previous module should be replaced by a parametric approach for including the effect of exogenous variables (as, for example, GAM or VAR models).

1.2.2 Case Study

This section contains the results of the generation of monthly precipitation fields for the main stations from Dobrogea, for the period Jan. 1966–Dec. 2005. The simulation has been performed using MMF, implemented in SCL software [18] and AAP, implemented in Castalia [17].

Table 12 contains the tolerances used in the rainfall generation using MMF. The result of annual precipitation generation at Constanta is represented in Fig. 5 and that of three monthly precipitation series (Jan, Jun, Sep), in Figs. 6, 7 and 8 (see also the values from Table 13).

From the charts of simulated monthly precipitation we remark that the simulated values approximate well the record data on the scale between 0 and 90 %.

In this approach the variance has not been preserved, so the precipitation generation has also been done based on a bi-stadial scheme [19]. Its author emphasized

Table 12 Tolerances utilized in monthly rainfall field generation

Statistics	Tolerance
Mean (%)	7.5
Standard deviation (%)	7.5
Skewness	0.75
1st order correlation coefficient	0.2
Maximum, minimum, 2, 3-years low sum (%)	10

Fig. 5 Annual precipitation generated for Constanta series

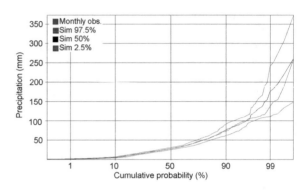

Fig. 6 Monthly precipitation —January—generated for Constanta series

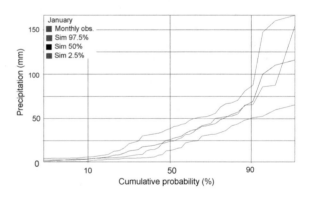

Fig. 7 Monthly precipitation —June—generated for Constanta series

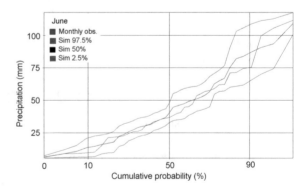

Fig. 8 Monthly precipitation —September—generated for Constanta series

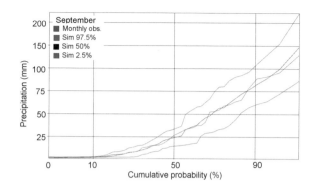

Table 13 Monthly precipitation generation by MMF at Sulina

(a) Annual data

	Record	Mean	2.50 %	50 %	97.50 %	Yes/No
Mean	261.634	270.176	216.021	267.513	334.153	Yes
Std. dev.	78.215	85.406	50.169	80.655	172.49	No
Skew	0.43	0.453	−0.592	0.402	1.742	Yes
1st order autocorrel. coef.	0.359	0.33	−0.047	0.34	0.685	Yes
Max	1.861	1.801	1.372	1.727	2.791	Yes
Min	0.419	0.432	0.131	0.449	0.642	Yes
2-years min sum	1.043	1.071	0.521	1.094	1.443	Yes
3-years min sum	1.759	1.809	1.083	1.836	2.365	Yes

(b) Monthly

Month	Record	Mean	2.50 %	50 %	97.50 %	Yes/No
Mean precipitation						
Jan	17.193	17.245	13.298	16.904	23.587	Yes
Feb	16.937	17.171	13.723	17.014	22.173	Yes
Mar	15.883	16.561	11.378	16.576	21.958	Yes
Apr	17.022	17.24	12.916	17.13	21.368	Yes
May	22.163	20.945	14.34	20.488	28.668	Yes
Jun	28.988	31.087	24.469	30.86	40.929	Yes
Jul	23.512	21.227	15.425	21.126	27.496	No
Aug	29.01	36.007	21.069	33.838	61.716	No
Sep	29.366	31.98	16.055	31.364	47.675	No
Oct	16.176	18.147	13.518	17.907	24.423	No
Nov	23.471	24.724	16.763	24.001	33.866	Yes
Dec	21.915	17.842	12.571	17.40	26.518	No

(continued)

Table 13 (continued)

(b) Monthly

Month	Record	Mean	2.50 %	50 %	97.50 %	Yes/No
Standard deviation						
Jan	14.996	13.358	866.70	1233	3008.10	No
Feb	13.619	14.297	11.019	14.326	19.074	Yes
Mar	14.36	15.903	11.049	16.117	19.797	No
Apr	12.778	12.509	8.005	12.789	18.953	Yes
May	18.069	14.696	9.212	12.659	28.326	No
Jun	16.853	18.401	13.942	18.23	23.823	No
Jul	21.143	18.319	13.014	18.13	23.031	No
Aug	27.695	34.559	18.07	34.164	56.195	No
Sep	29.837	31.472	18.37	31.82	42.326	Yes
Oct	12.813	16.123	11.09	16.135	20.924	No
Nov	19.913	20.446	11.629	20.448	28.834	Yes
Dec	18.122	14.171	8.423	13.729	30.999	No
Skewness coefficient						
Jan	1.704	1.026	−0.149	0.981	2.58	Yes
Feb	1.515	1.412	0.544	1.450	2.00	Yes
Mar	1.749	1.437	0.602	1.453	2.356	Yes
Apr	2.141	1.537	0.094	1.884	3.036	Yes
May	1.516	1.106	−0.193	0.727	3.253	Yes
Jun	0.29	0.382	−0.13	0.379	0.851	Yes
Jul	1.327	1.354	0.574	1.34	2.254	Yes
Aug	1.402	1.385	0.304	1.41	2.613	Yes
Sep	1.358	1.194	0.307	1.187	2.23	Yes
Oct	1.513	1.57	0.591	1.544	2.454	Yes
Nov	1.727	1.515	0.378	1.568	2.506	Yes
Dec	1.437	1.574	0.511	1.559	2.789	Yes
1st order autocorrelation coefficient						
Jan	0.19	0.073	−0.313	0.079	0.416	Yes
Feb	0.147	0.237	−0.036	0.249	0.623	Yes
Mar	−0.01	0.138	−0.046	0.13	0.333	Yes
Apr	−0.01	0.024	−0.244	0.018	0.323	Yes
May	0.176	0.274	0.042	0.253	0.587	Yes
Jun	0.108	0.231	−0.062	0.236	0.521	Yes
Jul	−0.22	−0.169	−0.431	−0.165	0.191	Yes
Aug	0.048	0.022	−0.299	0.014	0.323	Yes
Sep	0.072	0.181	−0.348	0.20	0.60	Yes
Oct	−0.15	−0.003	−0.293	−0.021	0.415	Yes
Nov	−0.013	0.006	−0.318	0.013	0.288	Yes
Dec	−0.005	0.018	−0.259	0.013	0.386	Yes

(continued)

Table 13 (continued)

(b) Monthly

Month	Record	Mean	2.50 %	50 %	97.50 %	Yes/No
Maximum precipitation						
Jan	73.7	58.51	34.097	52.24	125.38	No
Feb	63.9	59.128	44.44	60.845	66.7	Yes
Mar	63.9	60.214	41.564	62.769	67.792	Yes
Apr	71.9	61.483	32.163	74.366	88.376	No
May	86.7	68.421	33.303	53.715	130.359	No
Jun	66.1	69.845	57.95	69.072	88.802	Yes
Jul	89.5	77.018	50.227	81.114	86.195	No
Aug	129	138.41	69.743	143.194	194.16	Yes
Sep	126.3	117.517	73.925	121.616	160.6	Yes
Oct	63.2	64.684	43.537	66.765	70.831	Yes
Nov	94.6	87.56	46.556	94.167	101.28	Yes
Dec	75.8	65.09	34.411	68.426	128.956	No
Minimum precipitation						
Jan	1.5	1.515	1.269	1.446	2.864	Yes
Feb	1.5	1.582	1.427	1.513	1.992	Yes
Mar	0.3	0.52	0.253	0.298	2.035	No
Apr	3.4	3.288	0.914	3.566	4.254	Yes
May	0.7	1.034	0.269	0.659	5.796	No
Jun	0	1.5	0	0	5.721	No
Jul	0	0.311	0	0	1.707	No
Aug	0	0.006	0	0	0	Yes
Sep	0.1	0.284	0.095	0.103	1.6	No
Oct	1.7	2.794	1.727	2.192	4.997	No
Nov	1	1.61	0.939	1.821	2.097	No
Dec	2.7	2.978	0.726	3.15	4.311	No
Rainfall percentage zero rain						
Jan	0	0	0	0	0	Yes
Feb	0	0	0	0	0	Yes
Mar	0	0	0	0	0	Yes
Apr	0	0	0	0	0	Yes
May	0	0	0	0	0	yes
Jun	2.439	2.268	0	2.439	7.317	Yes
Jul	2.439	3.073	0	2.439	9.756	Yes
Aug	9.756	12.707	2.439	12.195	24.39	Yes
Sep	0	0	0	0	0	Yes
Oct	0	0	0	0	0	Yes
Nov	0	0	0	0	0	Yes
Dec	0	0	0	0	0	Yes

the capacity of this algorithm of keeping the most important statistics of the historic time series (mean, variance, standard deviation, skewness coefficient), the first order autocorrelation coefficient, the long range dependence property and the periodicity.

Here we present the simulation results for Constanta series.

The stages were the following:

- *Statistical analysis of the data.* The mean, standard deviation, the first order autocorrelation coefficients and the skewness coefficients of the annual and monthly data have been computed. The LRD has been analysed, and the extremes' distribution has been detected. It is of Log-Pearson III type for the annual data (1965–2005).
- *Generation of annual precipitation series.* This has been done by a symmetric moving average model (SMA), whose equation is:

$$X_i = \sum_{j=-s}^{s} a_{|j|} V_{i+j},$$

where $\{X_i\}$ is the series and $\{V_i\}$ are the innovations, used for reproducing the persistence in time.

It is known that the relationship between the coefficients a_i and the annual autocovariance function at the lag i, γ_i, is

$$\sum_{j=-s}^{s-i} a_{|j|} a_{|i+j|} = \gamma_i.$$

Moreover, taking into account the properties of annual series, the autocovariance function defined for generating the annual series is given by:

$$\gamma_h = \gamma_0 (1 + k\beta h)^{-1/\beta},$$

where h is the lag and k, and β are parameters. For Constanta annual series their estimated values are respectively 2.098 and 0.563.

The autocorrelation function estimated based on the previous formula and the annual standard deviations are presented in Fig. 9.

Fig. 9 Estimated annual autocorrelation function:
$\rho(h) = (1 + 1.181h)^{-1.776}$

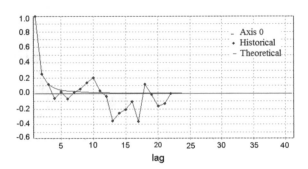

Table 14 Basic statistics of the generated series (for Constanta)

Month	Mean	Std. dev.	Skewness	1st order autocorrel. coef.
Jan	31.302	30.371	1.978	−0.052
Feb	26.454	21.204	1.762	−0.091
Mar	32.049	26.399	1.351	−0.046
Apr	30.407	20.627	1.130	0.054
May	36.310	28.581	1.372	0.221
Jun	42.907	26.163	0.891	0.066
Jul	31.478	23.278	1.960	−0.112
Aug	38.382	47.778	2.853	0.201
Sep	38.385	33.989	1.287	−0.185
Oct	32.970	24.216	1.957	0.074
Nov	43.797	30.542	0.824	0.075
Dec	36.408	31.997	2.973	−0.175
Annual	420.782	101.031	0.576	0.250

- *Generation of monthly series* by a seasonal autoregressive model PAR(1), that has the equation:

$$X^m = A^m X^{m-1} + B^m V^m,$$

where X^m, X^{m-1} are matrices that contain the monthly data, V^m contains the covariances, and A^m, B^m are the matrices of coefficients, depending on the mth month [19].

- *Computing the values of monthly series by disaggregation.*

The disaggregation procedure is used to assure the statistics' consistency at different temporal scales, the monthly series being adjusted at the annual ones.

For increasing the efficiency of this algorithm, Koutsoyiannis and Manetas proposed a Monte Carlo simulation scheme for each period. Data for many

Fig. 10 Monthly means of the simulated monthly series

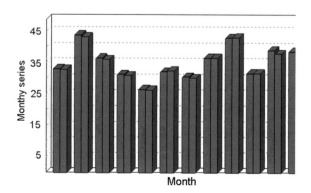

Fig. 11 Standard deviations
of the record and synthetic
monthly series

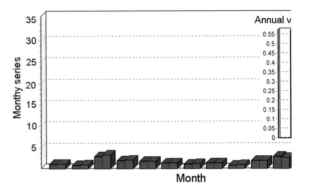

Fig. 12 Skewness
coefficients of the record and
synthetic monthly series

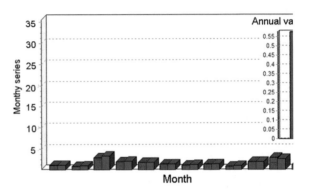

Fig. 13 First order
autocorrelation coefficients
for monthly series

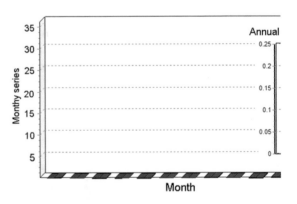

sub-periods are generated, by using PAR(1), until the minimization of the sum of
absolute values of the variance coefficients [20].

In the present application, 1000 synthetic series have been generated, using
Castalia software [17]. The mean values of basic statistics of the simulated series,
for Constanta, are presented in Table 14.

Table 15 Mean values of the statistics of 200 synthetic series generated for rainfall forecasting at Constanta for the period 2006–2015 (obtained using Castalia software)

	10	11	12	1	2	3	4	5	6	7	8	9	Mean	Stdev	Var coef	N	Miss	Max	Min	Up lim	Low lim	N Hi	N Low
2005-06	18.44	17.14	22.09	9.72	18.74	38.15	21.28	33.24	6.72	25.26	6.68	20.71	19.85	9.58	0.48	12	0	38.15	6.68	48.60	-8.90	0	0
2006-07	79.01	44.50	31.21	51.14	19.80	4.27	32.69	49.46	8.93	14.48	44.68	19.20	33.28	21.53	0.65	12	0	79.01	4.27	97.88	-31.32	0	0
2007-08	49.64	49.04	15.25	75.15	3.60	38.36	54.10	24.89	25.61	36.20	5.74	20.48	33.17	21.31	0.64	12	0	75.16	3.60	97.11	-30.76	0	0
2008-09	68.91	37.77	25.62	23.88	57.07	29.75	29.10	33.18	24.60	25.85	60.93	122.89	44.96	29.11	0.65	12	0	122.89	23.88	132.28	-42.36	0	0
2009-10	42.13	21.98	38.82	23.43	32.08	29.33	24.10	73.79	72.47	23.56	35.55	12.82	35.84	19.19	0.54	12	0	73.79	12.82	93.41	-21.73	0	0
2010-11	53.86	30.75	52.22	55.20	51.75	9.21	43.92	11.36	10.09	41.63	13.76	112.71	40.54	29.31	0.72	12	0	112.71	9.21	128.46	-47.39	0	0
2011-12	16.53	59.59	190.00	11.77	22.22	30.91	23.57	83.80	130.33	27.37	27.56	124.95	62.38	57.60	0.92	12	0	190.00	11.77	235.18	-110.42	0	0
2012-13	21.40	58.14	16.45	16.95	29.83	55.75	13.69	86.37	61.65	40.13	35.85	32.87	39.09	22.32	0.57	12	0	86.37	13.69	106.06	-27.88	0	0
2013-14	51.09	34.01	20.62	35.36	19.81	37.52	45.22	12.87	55.53	45.11	17.86	158.04	44.42	38.36	0.86	12	0	158.04	12.87	159.51	-70.67	0	0
2014-15	50.11	88.71	8.45	11.46	4.29	125.93	21.33	51.36	25.35	18.33	36.58	16.70	38.22	36.46	0.95	12	0	125.93	4.29	147.59	-71.16	0	0
Mean	45.11	44.16	42.07	31.41	25.92	39.92	30.90	46.03	42.13	29.79	28.52	64.14											
Standard deviation	21.01	21.05	53.49	22.28	17.63	33.62	12.91	27.77	38.74	10.38	17.66	57.74											
Variance coefficient	0.47	0.48	1.27	0.71	0.68	0.84	0.42	0.60	0.92	0.35	0.62	0.90											
Number of values	10	10	10	10	10	10	10	10	10	10	10	10											
Missing values	0	0	0	0	0	0	0	0	0	0	0	0											
Maximum value	79.01	88.71	190.00	75.16	57.07	125.93	54.10	86.37	130.33	45.11	60.93	158.04											
Minimum value	16.53	17.14	8.45	9.72	3.60	4.27	13.69	11.36	6.72	14.48	5.74	12.82											
Upper limit	108.13	107.31	202.54	98.26	78.81	140.77	69.62	129.35	158.34	60.92	81.51	237.35											
Lower limit	-17.91	-18.99	-118.40	-35.44	-26.97	-60.93	-7.82	-37.29	-74.08	-1.34	-24.46	-109.07											
High values	0	0	0	0	0	0	0	0	0	0	0	0											
Low values	0	0	0	0	0	0	0	0	0	0	0	0											

Fig. 14 Charts of the
monthly values for different
non-exceedance probabilities

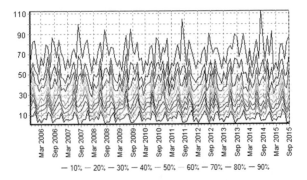

Fig. 15 Scenario of annual
rainfall prediction at
Constanta

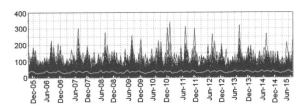

Fig. 16 Scenario of monthly
rainfall forecast at Constanta

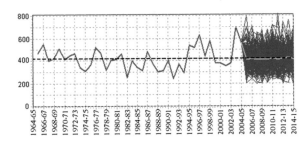

Fig. 17 Equal-probability
curves at Constanta, for
different exceedance levels

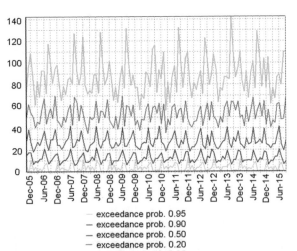

Table 16 Record monthly rainfall at Constanta in the period Oct 2005–Sep 2009

Month	Oct	Nov	Dec	Jan	Feb	Mar	Apr	May	Jun	Jul	Aug	Sep
2005–2006	64.4	93.9	26	20	31.8	68.4	25.4	82.6	12.4	76.6	50.6	46.6
2006–2007	4	27.8	12	28.1	16.4	35.2	21.2	19.2	21.4	4	84.4	54.8
2007–2008	53.4	108	48	39.8	0.5	24.2	109	52.8	50.2	27.4	1.8	20.6
2008–2009	9.6	39.4	49	37.8	36.6	14.4	24.7	49.1	4.4	107	5.8	44.6

Figures 10, 11, 12 and 13 contain comparisons of some statistics of the generated and historical series.

We used the obtained model for forecasting the precipitation evolution at Constanta for the period 2006–2015, generating 200 synthetic series. The average values of the statistics are presented in Table 15.

The charts containing the monthly values corresponding the non-exceedance probabilities of 5, 20, 50, 80 and 95 % are presented in Fig. 14.

The result of scenario of annual and monthly prediction of rainfall evolution is contained, respectively in Figs. 15 and 16, where the record data is in red, the generated one, in blue and the average forecast scenario in green.

Higher the number of scenarios is, the curve representing the average scenario tends to stabilize about the average annual value of the historic data.

The charts of the curve of equal probabilities for annual data are found in Fig. 17.

Remark that even if the generation of the precipitation fields at annual and monthly scale gave good results, the prediction scenario didn't give expected results, at least for the first data for which the record are available (Table 16).

This finding confirms our analysis concerning the absence of the temporal correlation of precipitation data in Dobrogea, that will be exposed in the next chapter.

2 Modeling and Forecast Using GEP, AdaGEP, SVR, GRNN and Hybrid Models

Recent studies proved that the use of modern modeling techniques instead of the traditional deterministic ones results in improving the knowledge on the evolution and forecasting meteorological time series.

For predicting the future evolution of any series, the first step is building a good model that accurately describes the past behavior of the process. Following this aim, in this section we analyze what technique could be more competent for modeling precipitation series from Dobrogea. In this way, we continue our previous work in this direction [1, 9, 10], comparing the performances of GEP, AdaGEP, SVR, and GRNN.

For utilizing GEP and AdaGEP modeling, each series has been preprocessed, resulting in w-dimensional vectors $(x_{t-w}, \ldots, x_{t-1}, x_t)$. When dealing with monthly series, that present seasonality, the input data used for the estimation of x_t are the values registered in the previous month, together with those from the same month of the previous year, i.e. $x_t = f(x_{t-1}, x_{t-12})$.

When using daily or annual data, different windows, w, are considered.

Generally, the data series is divided in two different sets: the first one, for training the algorithm, and the second one, for the validation of the model's capacity to predict the future values. Different authors used different ratios between the numbers of elements in the first set and in the second one, as, for example, 60:40, 70:30 or 90:10 [8, 14]. In the following, we shall mention the number of data used for validation for each study case.

Selecting the best individuals that survive in the process is done based on the fitness function, that was chosen to be MSE (mean squared errors); therefore the individuals that survive are that with the smaller MSEs.

Here are the parameters' set-up in our experiments with GEP and AdaGEP [14]:

- Gene length: 10 symbols and the genehad—5 symbols;
- The set of function: $\{+, -, \times, /, \sin\}$;
- The mutation rate: 0.03;
- The IS and RIS transpositions rates: 0.1;
- The one-point crossover and two-points crossover rate: 0.3;
- The gene-crossover rate and the transposition rate: 0.1;
- The linking function: addition;
- The selection scheme: the roulette wheel;
- The set of terminal values, $x_1, \ldots, x_{t-w}$, was determined function of the data series;
- The total number of genes in a chromosome was set to 5, but during the evolution process AdaGEP algorithm searched the optimal number of active genes;
- The genemaps mutation rate: 0.01;
- The genemaps crossover rate: 0.65.
- The number of the independent run for each experiment: 50.

Here we present only the best solution detected in the experiments.

The validation set has been formed by data recorded respectively in the last 24 month, or 12 years, when the series modeled were monthly series, respectively yearly.

To apply the ε-SVR, the RBF kernel has been used. It is defined by:

$$K(x_i, x_j) = \exp\left\{-\gamma \|x_i - x_j\|^2\right\},$$

Table 17 Results of modeling the precipitation series on the training dataset

Series	Method	RMSE	MAPE	Correlation actual-predicted
AMP	GEP	31.61	255.83	0.1827
	AdaGEP	30.34	240.23	0.147
	SVM	31.45	259.52	0.161
	GRNN	1.38	0.06	0.999
CMP	GEP	28.98	258.17	0.1334
	AdaGEP	38.19	3	0.5714
	SVM	28.17	267.22	0.2373
	GRNN	2.54	5.55	0.9861
MAP	GEP	64.9	11.78	0.6077
	AdaGEP	59.59	13.38	0.5478
	SVM	85.29	14.29	0.5153
	GRNN	80.36	13.69	0.6898
SMP	GEP	18.99	265.94	0.0556
	AdaGEP	28.54	2.16	0.8971
	SVM	19.31	209.62	0.2561
	GRNN	18.08	244.33	0.3947

Table 18 Results of modeling precipitation series on the validation dataset

Series	Method	RMSE	MAPE	Correlation actual-predicted
AMP	GEP	32.94	94.53	−0.096
	AdaGEP	29.45	75.22	0.367
	SVR	33.77	96.98	0.103
	GRNN	33.02	91.68	0.109
CMP	GEP	29.09	528.66	0.1969
	AdaGEP	6.53	2.88	0.782
	SVR	30	594.75	0.0491
	GRNN	29.78	580.34	0.0783
MAP	GEP	72.56	14.68	0.6444
	AdaGEP	72.34	14.7	0.6423
	SVR	103.34	20.71	0.6103
	GRNN	107.21	19.9	0.6471
SMP	GEP	10.47	270.97	0.2360
	AdaGEP	46.93	1.15	0.1251
	SVR	9.85	197.5	0.2508
	GRNN	10.25	247.09	0.2067

where $\|x_i - x_j\|$ is the distance between x_i, x_j and γ is a parameter.

The parameters' choice has been done by 10-fold cross-validation.

To compare the models' performances, the root mean squared error (RMSE) and the mean absolute percentage error (MAPE) have been used.

Remember that:

$$RMSE = \sqrt{\frac{\sum_{t=1}^{n} (x_t - x_t^*)^2}{n}},$$

and

$$MAPE = \sum_{t=1}^{n} \frac{|x_t - x_t^*|}{x_t} \times 100,$$

where x_t is the recorded value, x_t^* is the estimated one and n is the number of values.

These two indicators have been employed due to their different properties: RMSE depends on the scale and is affected by the highest errors in the model, but MAPE is independent on the scale, so it is a better measure of model's performances.

Closer to zero MAPE is, better the model is.

The studied series are annual and monthly precipitation series, as follows:

- Medgidia annual series, registered in the period 1965–2005 and denoted respectively by MAP;
- Adamclisi monthly series, registered in the period 1965–2005, denoted by AMP;
- Constanta monthly series recorded in the period 1961–2009, denoted by CMP;
- Sulina monthly series registered in the period 1961–2007, denoted SMP.

Tables 17 and 18 contain the values of RMSE, MAPE and the correlation between the actual and predicted values for the models of the studied series.

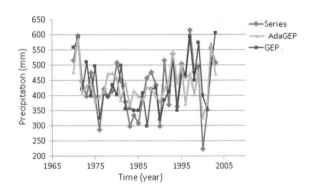

Fig. 18 GEP and AdaGEP for MAP with a window size $w = 5$

In the following, we discuss the performances of the algorithms on the analyzed data series.

(a) *Performances of the algorithms with respect to RMSE*

On the training sets the best results have been obtained using GRNN, in three out of four cases, i.e. for AMP, CMP and SMP. AdaGEP was more performant on the training and validation sets for MAP.

On the validation sets, AdaGEP gave better results in three out of four cases, especially for CMP. On this set, SVM performed better.

GEP and AdaGEP gave comparable results on MAP and AMP, but GEP was more competent on CMP and SMP.

For exmple, for MAP, AdaGEP was more performant than GEP as average number of symbols used over all the runs: AdaGEP used 20 symbols and GEP, 34. Also, the average genes' number in the best-of-run solutions of AdaGEP was smaller than for GEP: 3.5, in AdaGEP and 5, in GEP. Ada GEP has registered analogous performances on other time series [2].

Overall, on both series of sets, AdaGEP was the most performant (four cases), followed by GRNN (three cases) and SVR (one case).

Figure 18 contains the charts of MAP together with the models obtained using GEP and AdaGEP.

Remark that the shape of the AdaGEP chart is similar to that of the data series.

(b) *Performances of the algorithms with respect to MAPE*

The values of MAPE on the training sets shows that GRNN learns better the data than the other algorithms, and is on the second place as performances on the validation data sets.

On the training datasets, AdaGEP gave the best results in two out of four cases, and on the validation datasets, it performed better in three out of four situations. So, overall, AdaGEP was the best method for modeling the study data.

(c) *Performances of the algorithms with respect to the correlation between the actual and predicted values.*

The last columns in Tables 17 and 18 contain the correlations between actual and predicted values on the training and validation series. It can be seen that the most competent algorithm on the training set was GRNN. It learned well the data, but it overfits it on AMP (Fig. 19).

Fig. 19 GRNN model for AMP on the training set (obtained using DTREG software)

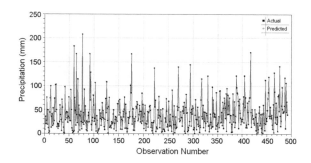

On the validation sets, the best performances have been obtained by using AdaGEP.

The Kolmogorov-Smirnov test has been used as a goodness of fitness test, to test the hypothesis that the registered and estimated data come from the same population [15]. The null hypothesis could not be rejected for all the models, but for those for SMP obtained by applying GEP and AdaGEP. Thus, GEP and AdaGEP models for SMP do not estimate correctly the distribution of the original data.

The monthly series are long, present seasonality and multiple break points. Therefore, better results are obtained on the sub-series determined by the change points or after the seasonality removal.

For example, in [9] we compared the performances of ARMA and AdaGEP models built for Medgidia monthly series (1965–2005). Since the series' skewness coefficient is significant, the series has been firstly transformed to reach the normality, using a Box–Cox transformation, with the parameter $\lambda = 0.39$. Then, the seasonality factors (respectively 79.7, 82.6, 83.9, 97.6, 111.9, 130.1, 126.6, 106.5, 96.4, 90.2, 97.6, 96.9) have been removed and an AdaGEP model has been built, using a window size of 12 (Fig. 20). The RMSE associated with it was of 10.04, but, still, the model obtained was worse than an ARMA(2, 2) [9].

The segmentation algorithm of Hubert applied to MMP detected two break points: one at the 76 data and the second one at the 319 data. Therefore, the series has been divided in three sub-series, S1—from January 1965 to the first breakpoint, S2—between the breakpoints, S3—from the second breakpoint to the end of the series.

The best model found for S1, after the Box-Cox transformation with the parameter $\lambda = 0.39$ was of MA(2) type, with a RMSE of 27. For S2, the best model was of AdaGEP type (with RSME = 26.13) and for S3, of MA(11) type.

We remark that even if methods coming from artificial intelligence are good alternatives for modeling time series that present high variability, they do not conduct always to better results than the classical methods.

The performances of each algorithm are highly dependent on the data, the structure and size of the training set.

Fig. 20 AdaGEP model: actual and predicted values on validation data set for Medgidia monthly series after a Box-Cox transformation and the seasonality removal

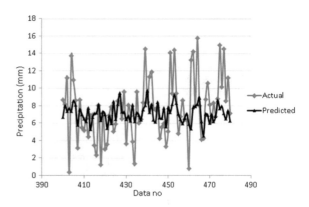

One of the possible causes of low performances of SVR might be the use of a single kernel which is not capable of generating accurate data estimations. Therefore, a general frame for building optimal complex kernels for Support Vector Classification has been introduced in [32], and tested with promising results.

Creating optimal multiple SVR kernels and their use for time series prediction has been also considered in [8]. The multiple kernels proposed are built by many single Gaussian, polynomial or sigmoidal kernels interconnected by operations from the set: {+, *, *exp*}.

Building multiple kernels is done by an evolutionary method with two levels: (1) the macro-level, where the multiple kernels are coded into chromosomes, (2) the micro-level, where the chromosomes' quality is estimated using a SVR algorithm.

As in the case of the evolutionary algorithms, the data are divided in two parts—one for training and another for validation—with the aim of evaluating the chromosomes' quality. The difference between this algorithm and other known methods is that the training subsets is divided, at its turn, in two subsets—the learning one, for training the SVR and the validation one, for estimating the fitness function.

This algorithm has been used for forecasting financial time series and the results are presented in the article [31].

3 Decomposition and Wavelets Models

As previously discussed, decomposition models take into consideration the existence of the multiple components of a time series and offer the advantage of separately examining each compound.

Here we present the decomposition of Medgidia monthly series.

Statistical tests on this series proved that the series is not Gaussian (Fig. 21), is not autocorrelated and presents outliers. In the modeling process, the values registered in May 1971 and June 1991 have been removed, being outliers.

The homoskedasticity hypothesis could not be rejected.

The segmentation algorithm of Hubert determined two break points: April 1971 and July 1991. The Pettitt and Bai-Perron algorithms didn't reject the hypothesis that there is no break point in the time series. Therefore, models have been built for the entire series, denoted by S, and for its sub-series, denoted by S1, S2 and S3, respectively.

The trend of the study series has been determined as a polynomial of the sixth order, for S, S1, S2 and of third order for S3. In Figs. 22, 23, 24 and 25 we present the data and the trend. The seasonal factors are given in Table 19.

We remark that July was strongly affected by seasonal variations. High seasonality factors of S and S1 correspond to December, respectively May and July for S2 and S3.

After the residual analysis, it resulted that residuals are normally distributed, uncorrelated and homoskedastic, with the mean squared errors between and 1.3906 and 1.5507, in all models.

Fig. 21 Q-Q plot of
Medgidia monthly series

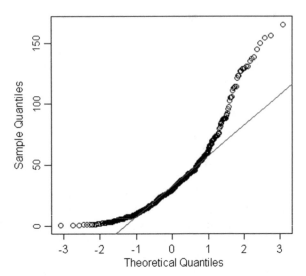

Multiplicative decomposition models have also been determined for Mangalia
and Sulina monthly precipitation series (1965–2005).

After deciding on the optimal segmentation, based on the results given by different methods (Pettitt, Hubert, mDP and Bai-Perron), Sulina series has been
normalized by a Box-Cox transformation with the parameter 0.34. Then, it has

Fig. 22 Medgidia monthly
series and the polynomial
trend

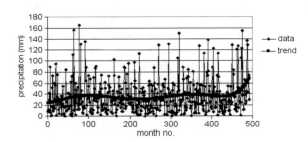

Fig. 23 S1 series and its
trend

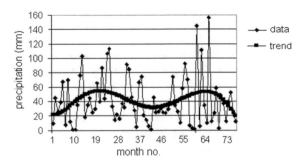

Fig. 24 S2 series and its trend

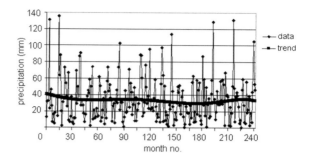

Fig. 25 S3 series and its trend

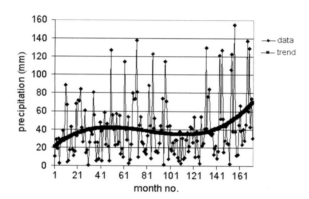

Table 19 Seasonal factors (polynomial trend)

Month	Jan 1965–Dec 2005	Jan 1965–Apr 1971	Jun 1971–May 1991	Jul 1991–Dec 2005
January	1.078419	1.11236	0.649563	0.5419365
February	1.215249	1.317954	0.6950985	0.4878041
March	0.665324	0.745415	0.6360678	0.8294957
April	0.542969	0.638693	1.014415	0.8229927
May	1.202965	1.185166	1.286115	1.0365083
June	0.742822	0.76237	1.7565319	1.526265
July	1.48271	1.493408	1.3664652	1.5882092
August	1.171261	1.117817	1.1633314	1.0610638
September	0.689192	0.651216	0.9058862	1.5380581
October	0.597726	0.581296	0.8200046	0.8991314
November	0.836525	0.767221	0.9823767	0.892835
December	1.774838	1.627086	0.7241447	0.7756996

been divided into sub-series that, in their turn, have been decomposed based on a multiplicative model. For example, for the entire series, the trend's equation is:

$$Y_t = 4.67 + 0.53 \cos(0.0077t - 0.36),$$

where t is the time.

The months most affected by seasonality were June and September [3].

The modeling results (for the entire series and its subseries) are comparable with respect to the mean standard errors.

For Mangalia, no change point has been detected, so the decomposition has been done only for the entire series, whose trend has been determined to be of cosine type. For details, see [3].

As mentioned, the degree of polynomial that describes the trend in the model for Medgidia series and its sub-series is high. Therefore it does not assure the model' stability. Thus other models for the series' evolution have been proposed.

Wavelets models of precipitation evolution for these series are presented in Figs. 26, 27, 28 and 29. They have been obtained using the hard thresholding.

Looking to Figs. 26, 27, 28 and 29 we remark the existence of some intervals of relative homogeneity of the precipitation quantity followed by periods of high rainfall.

High variability of precipitation record is observed in S2, for which the model describes very well the series evolution (Fig. 28).

MSEs of residuals in the wavelets models for S, S1-S3 are respectively: 31.83, 41.244, 34.287, 37.211, so the best model is that of S series.

Wavelets analysis can be also used to distinguish the trend, change points and extreme values in a time series. Saif et al. [28] proposed such an algorithm for financial time series analysis, which was then successfully applied for modeling precipitation series [25].

Fig. 26 Wavelets model for Medgidia monthly series (Jan. 1965–Dec. 2005)

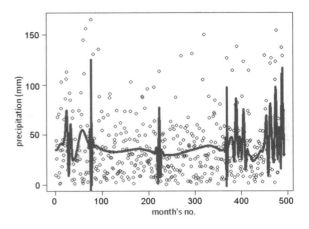

Fig. 27 Wavelets model for
Medgidia monthly sub-series
S1 (Jan. 1965–Apr. 1971)

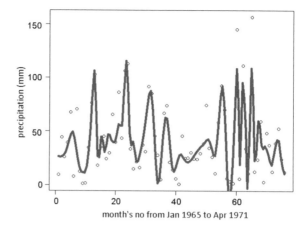

Fig. 28 Wavelets model for
Medgidia monthly sub-series
S2 (June 1971–July 1991)

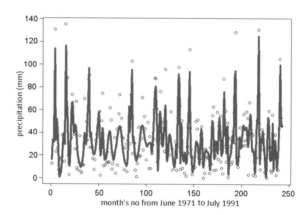

Fig. 29 Wavelets model for
Medgidia monthly sub-series
S3 (Sep. 1991–Dec. 2005)

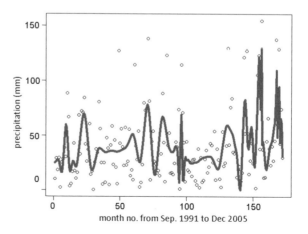

Fig. 30 Wavelets model for
annual precipitation evolution
at Ceamurlia (1965–2005)

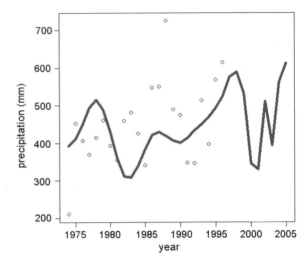

Fig. 31 Wavelets
decomposition for the winter
record of precipitation at
Constanta (1965–2005)

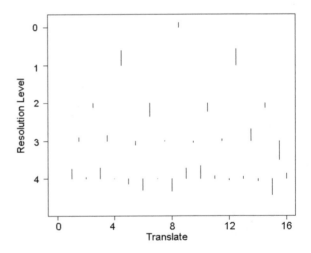

The algorithm has the following steps:

1. Data compression, for a given period (for example, 12 months), resulting in four
   series: B (Begin), *Ma* (Maximum), (*Mi*) Minimum and E (End);
2. Computation of the level, L, of the Discrete Wavelets Transform (DWT),
   function of the number of samples from the first step;
3. Performing a level-L DWT decomposition of E based on results of Steps 1 and
   2, for obtaining the details (D_i) at different levels (i = 1, ..., L) and highest-level
   approximation (A_L);
4. Computation of the trend by performing a linear regression on A_L;
5. Seasonality extraction by means of a Fourier power analysis on D_i;
6. Turning points extraction, by choosing extrema of each D_i.

In the case of Sulina and Medgidia monthly series, $N = 492$, $L = 6$, and $A_L = 8$.

The step 4 of the algorithm has been modified to permit the choice of a polynomial trend. For the study series, the results are concordant with those obtained by decomposition using the polynomial trend, i.e. the coefficients of the polynomial trend in both models coincide up to the sixth decimal.

Wavelets technique has been successfully used for modeling the evolution of precipitation for different series of annual [12], monthly or maximum precipitation at a site or in a region [4, 11]. In [13] we give the R code used for this purpose. Here, we analyze only some results of this approach.

In Fig. 30 we present the model for annual precipitation evolution in the period 1965–2005 at one of the secondary meteorological stations in Dobrogea, called Ceamurlia. The shape of the fitted series is similar to that of the historic data, with periods of higher precipitations, followed by periods of lower rainfall. Even if the rainfall is underestimated for the period 1982–1995, the standard error in the model is small—7.87 (mm).

Figure 31 contains the decomposition of the precipitation record in winter at Constanta station (1965–2005), in the same period, using Daubechies wavelets.

Figure 32 provides the smoothing data for winter record, obtained using the hard thresholding, and applied to different decomposition levels.

An increasing precipitation trend is observed for the threshold applied to levels 1:4. The models obtained applying the smoothing procedure at levels 3:4 relieve the periods of precipitation increasing and decreasing. The same shape can be observed for annual precipitation decomposition for Adamclisi (Figs. 33 and 34).

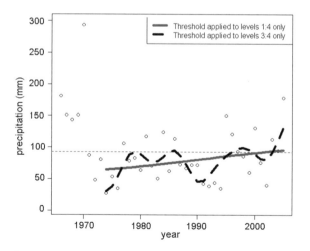

Fig. 32 Wavelets smoothing of precipitation record in winter at Constanta (1965–2005), using hard thresholding

Fig. 33 Wavelets
decomposition for the record
winter precipitation at
Adamclisi (1965–2005)

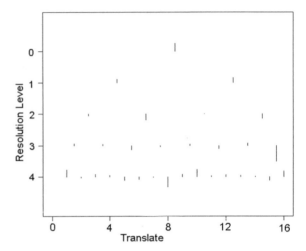

Fig. 34 Wavelets smoothing
of precipitation record in
winter at Adamclisi (1965–
2005), using hard
thresholding

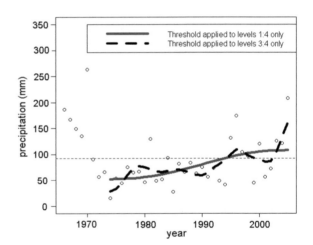

References

1. Bărbulescu, A., Băutu, E.: Mathematical models of climate evolution in Dobrudja. Theor. Appl. Climatol. **100**(1–2), 29–44 (2010)
2. Bărbulescu, A., Băutu, E.: Time series modeling using an adaptive gene expression programming. Int. J. Math. Models Methods Appl. Sci. **3**(2), 85–93 (2009)
3. Bărbulescu, A., Deguenon, J.: Change point detection and models for precipitation evolution. Case study, Rom. J. Phys. **59**(5–6), 590–600 (2014)
4. Bărbulescu, A., Deguenon, J.: Modeling the annual precipitation evolution in the region of Dobrudja. Int. J. Math. Models Methods Appl. Sci. **6**(5), 617–624 (2012)
5. Bărbulescu, A., Pelican, E.: ARIMA models for the analysis of the precipitation evolution. Recent Adv. Comput. 221–226 (2009)

6. Bărbulescu, A., Pelican, E.: On the Sulina precipitation data analysis using the ARMA models and a neural network technique. In: Mathematics and Computers in Science and Engineering, Part II, pp. 508–511 (2008)
7. Bărbulescu, A., Șerban (Gherghina), C., Maftei, C.: Statistical analysis and evaluation of Hurst coefficient for annual and monthly precipitation time series. WSEAS Trans. Math. **10**(9), 791–800 (2010)
8. Bărbulescu, A., Simian, D.: Theoretical and practical approaches for time series prediction. In: Proceedings of Third International Conference on Modeling and Development of Intelligent Systems, 10–12 Oct 2013, Sibiu, Romania, Lucian Blaga University Press, pp. 40–48 (2014)
9. Bărbulescu, A., Băutu, E.: Alternative models in precipitation analysis. Analele Științifice Universitatea Ovidius din Constanta, Matematica **17**(3), 45–68 (2009)
10. Bărbulescu, A., Băutu, E.: ARIMA and GEP models for climate variation. Int. J. Math. Comput. **3**(J09), 1–7 (2009)
11. Bărbulescu, A., Deguenon, J.: Models for trend of precipitation in Dobrudja. Environ. Eng. Manage. J. **13**(4), 873–880 (2014)
12. Bărbulescu, A., Petac, A.: Statistical assessement of precipitation evolution. Case study, Autom. Comput. Appl. Math. **22**(1), 7–15 (2013)
13. Bărbulescu, A., Deguenon, J., Teodorescu, D.: Study on Water Resources in the Black Sea Region. Nova Publishers, USA (2011)
14. Băutu, E., Bărbulescu, A.: Forecasting meteorological time series using soft computing methods: an empirical study. Appl. Math. Inf. Sci. **7**(4), 1297–1306 (2013)
15. Chakravarti, I.M., Laha, R.G., Roy, J.: Handbook of Methods of Applied Statistics, vol. I. Wiley, New York (1967)
16. https://cran.r-project.org/web/packages/moments/moments.pdf
17. https://www.itia.ntua.gr/en/docinfo/619/
18. http://www.toolkit.net.au/Tools/SCL/
19. Koutsoyiannis, D.: A generalized mathematical framework for stochastic simulation and forecast of hydrologic time series. Water Resour. Res. **36**(6), 1519–1533 (2000)
20. Koutsoyiannis, D., Manetas, A.: Simple disaggregation by accurate adjusting procedures. Water Resour. Res. **32**, 2105–2117 (1996)
21. Lane, W.L.: Applied stochastic techniques (LAST computer package), user manual. Division of Planning Technical Services, Bureau of Reclamation, Denver, Colorado (2000)
22. Lee, T., Salas, J.D., Prairie, J.: An enhanced nonparametric streamflow disaggregation model with genetic algorithm. Water Resour. Res. **46**, W08545 (2010). doi:10.1029/2009WR007761
23. Maheepala, S., Perera, C.J.C.: Monthly hydrologic data generation by disaggregation. J. Hydrol. **178**, 277–291 (1996)
24. Mejia, J.M., Rouselle, J.: Disaggregation models in hydrology revisited. Water Resour. Res. **12**, 185–186 (1976)
25. Pelican, E., Bărbulescu, A.: Forecasting rainfalls in Dobroudgea area using wavelets analysis, mathematical modeling of environmental and life sciences problems. In: Proceedings of the Sixth Workshop, Sept 2007. Proceedings of the Seventh Workshop, Sept, 2008, pp. 211–218. Constanța, Editura Academiei Romane, Bucuresti (2010)
26. Porter, J.W., Pink, B.J.: A method of synthetic fragments for disaggregation in stochastic data generation. Hydrology and Water Resources Symposium, Institution of Engineers, Australia, pp. 187–191 (1991)
27. Prairie, J., Rajagopalan, B., Lall, U., Fulp, T.: A stochastic nonparametric technique for space—time disaggregation of streamflows. Water Resour. Res. **43**, W03432 (2007). doi:10.1029/2005WR004721
28. Saif, A., Taskaya-Temizel, T., Khurshid, A.: Summarizing time series: learning patterns in volatile series. In: Yang, R., et al. (eds.) IDEAL 2004. Lecture Notes in Computer Science, vol. 3177, pp. 523–532. Springer, Berlin (2004)
29. Serinaldi, F., Kilsby, C.G.: A modular class of multisite monthly rainfall generators for water resource management and impact studies. J. Hydrol. **464–465**, 528–540 (2012)

30. Sharma, A., Mehrotra, R.: Rainfall generation, a review. In: Testik, F.Y., Gebremichael M. (eds.) Rainfall: State of the Science, pp. 215–246. AGU (2010)
31. Simian, D., Bǎrbulescu, A.: Financial Time Series Forecasting Using SVRs with Optimal Multiple Kernels (under review)
32. Simian, D., Stoica, F.: A general frame for building optimal multiple SVM kernels. In: Lirkov, I. et al. (eds.) Lecture Notes in Computer Science, vol. 7116, Large Scale Scientific Computation, pp. 256–263 (2012)
33. Srikanthan, R., McMahon, T.: Stochastic generation of annual, monthly and daily climate data: a review. Hydrol. Earth Syst. Sci. 5(4), 653–670 (2001)
34. Srikanthan, R., Chiew, F.: Stochastic models for generating annual, monthly and daily rainfall and climate data, at a site. Technical report 03/16, Cooperative Research Centre for Catchment Hydrology (2003)
35. Srikanthan, R., Kuczera, G., Thyer, M., McMahon, T.: Stochastic generation of annual rainfall data. Technical report 02/6, Cooperative Research Centre for Catchment Hydrology (2002)
36. Srikanthan, R., McMahon, T.: Stochastic generation of climate data: a review. Technical report 00/16, Cooperative Research Centre for Catchment Hydrology (2000)
37. Srikanthan, R., McMahon, T.: Stochastic generation of rainfall and evaporation data. AWRC technical paper no. 84, 301pp (1985)
38. Srikanthan, R., McMahon, T.A., Pegram, G.G.S., Kuczera, G.A., Thyer, M.A.: Generation of annual rainfall data for Australian stations. Working document 02/3, Cooperative Research Centre for Catchment Hydrology (2002)
39. Srikanthan, R., McMahon, T.A., Sharma, A.: Stochastic generation of monthly rainfall data. Technical report 02/8, Cooperative Research Centre for Catchment Hydrology. http://www.toolkit.net.au/Tools/SCL/publications (2002)
40. Stedinger, J.R., Pel, D., Cohn, T.A.: A condensed disaggregation model for incorporating parameter uncertainty into monthly reservoir simulations. Water Resour. Res. 21, 665–675 (1985)
41. Tarboton, D.G., Sharma, A., Lall, U.: Disaggregation procedures for stochastic hydrology based on nonparametric density estimation. Water Resour. Res. 34, 107–119 (1998)
42. Thomas, H.A., Fiering, M.B.: Mathematical synthesis of streamflow sequences for the analysis of river basins by simulation. In: Maass, A., et al. (eds.) Design of Water Resource Systems, pp. 459–493. McMillan, London (1962)
43. Thompson, C.S.: Homogeneity analysis of rainfall series: an application of the use of a realistic rainfall model. J. Clim. 4, 609–619 (1984)
44. Thyer, M., Kuczera, G.: Modeling long-term persistence in hydroclimatic time series using a hidden state Markov Model. Water Resour. Res. 36(11), 3301–3310 (2000)
45. Unal, N.E., Aksoy, H., Akar, T.: Annual and monthly rainfall data generation schemes. Stoch. Environ. Res. Risk Assess. 18, 245–257 (2004)

Chapter 4
Modeling the Precipitation Evolution at Regional Scale

The first part of this chapter consists of the approach of the independence of data series recorded at stations in Dobrogea and their predictability, using only historical data. Some methods used belong to the fuzzy techniques. Our conclusion is that the monthly precipitation series are not temporal correlated. Therefore, it is difficult to sustain the idea that the evolution of regional precipitation is predictable based only on the historical data.

The second part of the chapter contains the results of modeling the regional annual, monthly and monthly maximum annual rainfall series using different techniques. Splines and wavelets are utilized along with grouping the data in two different ways.

The study has practical implications. Given that the regional models built using homogeneous groups of series do not lead to loss of essential information, in the absence of some data they can be successfully used for studying the regional rainfall variability. Secondly, knowing the regional model, the missing data of a particular series can be replaced with those from the regional model of the homogeneous group to whom the series belongs.

1 On the Correlation and Predictability of Precipitation at Regional Scale

In this section, we study the possible predictability of the precipitation data using only their history, known that the subject is still controversial.

Some authors have proposed methods for generation possible scenarios for the future evolution of precipitation using only the record data and applied them successfully to particular regions [16, 17]. Other scientists [21] pointed out that chances to predict the long-term evolution of meteorological data are reduced due to the chaotic nature of climate system [20].

© Springer International Publishing Switzerland 2016
A. Bărbulescu, *Studies on Time Series Applications in Environmental Sciences*, Intelligent Systems Reference Library 103, DOI 10.1007/978-3-319-30436-6_4

Many approaches to predict the precipitation are based on numerical weather prediction (NWP) models [15], Bayesian methods [14, 29], the use of satellite or radar data [5, 34]. Wu et al. [35] pointed out that NWP models underperform for precipitation prediction. Duan et al. [8] compared CMA, ECMWF, NCEP and JMA models and emphasized that a proper behavior of a system is the result of four components: the system of data assimilation, the numerical models, the initial perturbations, as well as the model of stochastic variations.

In Chapter I.3 we discussed the independence test introduced in [32], denoted in the following by SRB test. Here we apply it to the dataset formed by the long series of monthly precipitation, recorded without gaps, at 49 meteorological stations from Dobrogea, in the period 1965–2004.

Here we explain some terms employed in the following:

- *The local precipitation data* is a time series record at a particular meteorological station;
- *The regional time-series* or *regional precipitation* data is a matrix collecting all the local time-series (as columns);
- *Self-predictability* of an n-dimensional time-series ($n \geq 1$) refers the situation in which the current dynamic of the time series is predictable from the series history;
- *The future sequence* (F) and *the historical data* (H) are relative with respect to the time: any current non-terminal reading of the time-series determines a pair (H_t, F_t).

The results of preliminary statistical analysis are summarized as follows:

- The monthly precipitation values range between 0 and 406.5 L/m^2, with local averages between 21.8 and 49.19 L/m^2 and standard deviations between 19.55 and 38.
- The normality tests rejected the null hypotheses for all the local series. Consequently, among the independence tests, only distribution-free tests can be used, as, for example, those proposed in [28, 32].
- The skewness and kurtosis coefficients are respectively in the intervals [1.22, 3.46] and [1.35, 23.45].
- The homoskedasticity tests didn't reject the null hypothesis for 26 series;
- Autocorrelation tests revealed the absence of autocorrelation for seven series—weak first-order autocorrelation for 39 series and weak higher order (2–4, 6, 9–13) autocorrelation for other local time-series. The maximum values of all autocorrelation coefficients were calculated for lags between 1 and 144 months. All these values are small, so the autocorrelation is very weak, indeed.

SBR test has been chosen to test the independence of actual and historical data series because it is implemented in R and doesn't require sophisticated mathematical background to be understood. Also, it is distribution-free, and it is unbiased for samples with the volume greater than three.

The series of tests performed on the regional rainfall data D (formed by 480 monthly values ×49 stations) have been organized as follows:

- T1: Input: The matrix of regional data (RD), divided in History (H) and Future (F), that cover the entire period; Output: temporal EDCor values;
- T2: Input: Pairs of history and future data, increasingly longer, that cover the entire period; Output: temporal EDCor values calculated for these sequences;
- T3: Input: $(X_{\sigma_k}, F_{\sigma_k})$ randomly temporal partition of history and future data; Output: temporal EDCor values computed for these partitions;
- T4: Input: $(X_{\sigma_k}, Y_{\sigma_k})$ spatial partitions of regional data obtained by random division of the matrix of regional series in matrices containing k and $n - k$ columns; Output: spatial EDC or values for randomly spatial partition of data.

Table 1 contains the results of the test T1. The second row presents the results of the SRB test whereas the data is formed respectively by 240 months of history (H), followed by 240 months of future data (F).

The third row of Table 1 contains the results of SRB test when the history data was formed by the first 240 months and the future data, of 240 arbitrary uniformly distributed random future data artificially generated. Comparing these two lines, we remark that the degree of statistical dependence between the actual history and future sequences is weaker than the dependence between the actual history and an arbitrary uniformly distributed random future data.

The fourth line of Table 1 illustrates the meaning of the results previously obtained by comparison to the results of SRB tests undertaken for the actual history data and the reversed future data. The results are comparable to those for History to Future test.

The test T2 illustrates the convergence of the EDCor coefficients computed along 20 increasingly longer history and future sequences (H_i, F_i), when i ranges from 1 to 20 and defines the sequences length, $12i$.

Here we remark the disadvantage of SRB test that is sensitive to the length of the data series, providing greater EDCor values for shorter data sequences So, these values are not relevant for the correlation coefficients for the entire series and long sub-series.

If there is the chance to predict future from history, then, a temporal permutation of the regional precipitation matrix should lead to important changes in the results

Table 1 EDCor values obtained in *T1*

Comparison	D
History to future	0.082356
History to random future	0.169570
History to reversed future	0.082007

of SRB tests. Therefore, the test T3 is designed as a Monte Carlo simulation in which EDCor is calculated for 240-long arbitrary history and future sequences selected by random permutation of data on the time axis.

In the test T4, the columns of regional precipitation matrix are randomly permutated in the spatial direction, then the matrix is partitioned into halves, on the same direction (24-24 stations, one station is excluded each time). For each such a transformation, EDCor is computed. This simulation, as well as that from T3, is repeated 5400 times.

The EDCor values for the temporal partition are around 0.1. Hence, there is no temporal dependence between these arbitrarily formed 'history' and 'future' sequences (Fig. 1, the black dots). The values of EDCor coefficients computed for the random halves of regional data in the 5400 simulations vary between 0.57 and 0.65 for the spatial partition (Fig. 1, the gray dots). Therefore, the hypotheses of statistical independence formulated for the randomized halves cannot be validated through the SRB test.

Independence tests have been performed on each series: (a) uniformly scaled down to the uint8 domain, (b) directly truncated to the uint8 domain, (c) truncated to the uint8 domain and fuzzified in five values (dry, normal, moist, wet and damp), (d) truncated to the uint8 domain and fuzzified in four values (dry, normal, moist and wet) (Table 2 and Fig. 2).

The test has been performed on actual data and the data truncated to the uint8 domain and for fuzzified 5/4-means data series. Each data set was divided into halves and the EDCor have been calculated.

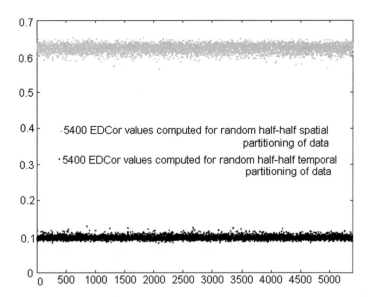

Fig. 1 T3 and T4 tests—The EDCor values computed for 5400 ad hoc random half-half spatial (*gray dots*) and temporal (*black dots*) partitioning of data (nos. of repetitions on *abscise*, EDCor values on *ordinate*)

Table 2 Centroids and labels in: **a** 4-label fuzzification and **b** 5-label fuzzification

(a)	Label	Dry	Normal	Moist	Wet	
	L/m²	10	30	52	75	
(b)	Label	Dry	Normal	Moist	Wet	Damp
	L/m²	10	30	51	75	100

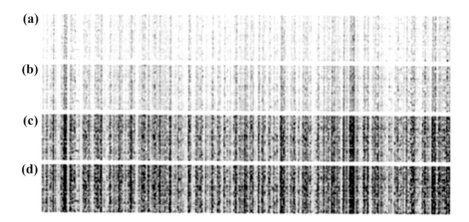

(a)

(b)

(c)

(d)

Fig. 2 Regional precipitation data: **a** uniformly scaled down to the uint8 domain, **b** directly truncated to the uint8 domain, **c** truncated to the uint8 domain and fuzzified in five values, **d** truncated to the uint8 domain and fuzzified in four values

Figure 3 illustrates the values of coefficients computed for the four different kinds of data presented in Fig. 2.

Figure 4 shows the convergence of the EDCor coefficients calculated for increasingly (one word) longer local historical and test data originating in the 1st

Fig. 3 T1—Testing for EDCor values: EDCor coefficients computed for the actual/truncated/fuzzified local data

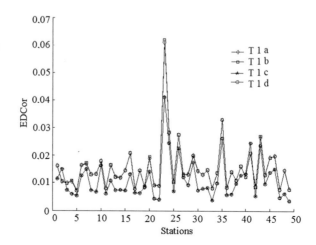

Fig. 4 The convergence of
EDCor coefficients computed
in *T1* for all the 49 local
time-series with the data
truncated to uint8 domain and
fuzzified in four values

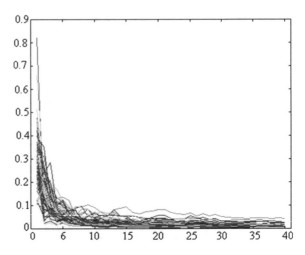

and 240th month, truncated to uint8 domain and fuzzified in four values. EDCor
tends to accumulate to zero, after a short period.

An extensive study of the convergence of EDCor on the fuzzified local and
regional precipitation data in Dobrogea appears in [23]. The results are concordant
with those on the raw data. Thus, the fuzzified data can be successfully used for the
study of the series' independence when the series contains gaps.

The study proves that the monthly precipitation data analyzed here doesn't present patterns at temporal scale (expressed in high correlation coefficients) that could
sustain the predictability of regional monthly precipitation data from their history.

2 Modeling the Regional Precipitation

The idea of modeling the regional precipititation series is not new, some aspects
being previously treated in [2, 4]. The spatial correlation of the precipitation record
data in Dobrogea in the period 1965–2005 sustains this atempt.

Here we summarise some results, and we present the model for the evolution of
maximum annual precipitation in Dobrogea.

Different approaches for modeling the regional precipitation are noticed. Some
of them involve the use of nonparametric methods, due to their capabilities of
describing the evolution of the study process whithout imposing restrictive
hypotheses on the input data or specifying in advance the type of relationship
between the endogenous and exogenous variables. Examples of these techniques
are: nonparametric regression, local polynomial smoothing, splines, etc. [9].

In the *local polynomial smoothing* [9], a polynomial is fit in a neighbourhood of
each point, using the weighted least squares. The weights are inverse proportional
to the distances between the point whose value is estimated and the other points. In
the case of the local linear smoothing, the order of the local polynomial is one.

To obtain a reasonable model for the regional data, the series' group must be homogenous. For testing the homogeneity in variance (homoskedasticity), the Levene test is performed. For testing the mean equality, different options are available. One of them is ANOVA, that is based on the following hypotheses:

(a) The distribution of all populations involved is Gaussian or approximately Gaussian;
(b) The populations' variances are equal;
(c) The samples are independent on each other and randomly selected.

ANOVA is robust to the normality violation when each group has the same number of observations, but is not robust to the violation of (b) and (c). Therefore, in such situations, the nonparametric test of Kruskal–Wallis [18] can be employed instead of ANOVA.

Since for the annual regional precipitation data all the hypotheses of ANOVA are satisfied, the Scheffé test has been performed, at the significance level of 0.01, as a confirmatory test of homogeneity. It resulted that Sulina series doesn't belong to the group formed by all the other series (Table 3).

At the significance level of 0.05, the following groups of data resulted:

- Sulina;
- *Group* I: Jurilovca, Harsova, Constanta, Mangalia, Corugea, Tulcea, Medgidia;
- *Group* II: Harsova, Constanta, Mangalia, Corugea, Tulcea, Medgidia, Adamclisi, Cernavoda.

Since the monthly precipitation series and the maximum monthly annual precipitation series are not Gaussian, the Kruskal–Wallis test has been performed for them. The test results rejected the null hypothesis in both situations. After repeated applications of the this test, the same homogenous groups of monthly precipitation series as for the annual data have been obtained.

Table 3 Results of the Scheffé test at the significance level $\alpha = 0.01$

Series	Groups	
	1	2
Sulina	261.63	
Jurilovca	378.39	378.39
Harsova		408.82
Constanta		423.04
Mangalia		427.74
Corugea		434.66
Tulcea		434.66
Medgidia		449.92
Adamclisi		484.54
Cernavoda		487.59
p-value	0.012	0.029

2.1 Models for the Regional Annual and Monthly Precipitation Series

For the regional precipitation series, different models can be built, using R software as an instrument.

In [1] we present the result of modeling the annual datasets (regional and without Sulina) by the local linear smoothing with the optimal bandwidth of 0.6988 (chosen by the general cross-validation principle) and the spline smoothing, with the default optimal bandwidth, the (computed) smoothing parameter 0.117317 and the number of knots K = 34.

The results of both methods are comparable with respect to the standard errors.

Using the dataset without Sulina results in a diminishing of the mean standard error with about 22 % (Table 4).

Figure 5 contains the record annual precipitation (as dots) and the fitted curves, denoted by local linear and splines. In this figure, the curves are practically superimposed.

Let denote by:

- K—the station's number,
- N—the number of values registered at each station, given that the data is recorded at the same time at all the stations and is without gaps,
- $Y = (Y_{11}, \ldots, Y_{1n}, \ldots, Y_{k1}, \ldots, Y_{kn})$—the regional precipitation vector, whose component Y_{ji} is the precipitation record at the j-th meteorological station, at the time t_i, $i = 1, \ldots, n$, $j = 1, \ldots, k$;
- $T = (t_1, \ldots, t_n)$—the vector of the moments.

Method A

- Build the vector $X = (X_1, \ldots, X_{kn})$, whose component $X_m, 1 \leq m < kn$ is t_n, if m is a multiple of n or t_r, where r is the remainder of the division of m by n, otherwise.
- Apply the modeling technique with X as independent variable and Y as the dependent one.

Table 4 Mean standard errors in local linear smoothing and splines smoothing for annual regional data

	Local linear	Splines
Regional dataset	87.06	87.64
Without Sulina	67.76	67.93

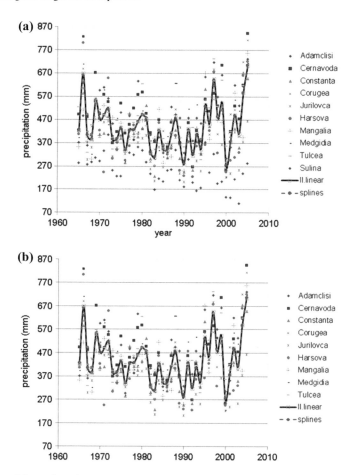

Fig. 5 Local linear fit and splines smoothing for the period 1965–2005, using: **a** annual regional precipitation; **b** the homogenous group of data

Method B

- Define $Z_i = \frac{1}{k}\sum_{j=1}^{k} Y_{ji}$, $i = 1, \ldots, n$, and $Z = (Z_1, \ldots, Z_n)$.

- Apply the modeling technique with T as independent variable and Z as the dependent one.

In the following, we discuss the splines and wavelets models for the annual regional dataset and the homogenous groups of series.

Using the splines smoothing technique for the annual data grouped as in A and B, the residual standard deviation in the models for Group I are respectively 86.85 and 64.65 and for Group II, 66.98 and 64.84.

Fig. 6 Wavelets model, using method *A*, for the annual regional precipitation (1965–2005): **a** without Sulina and **b** Group I

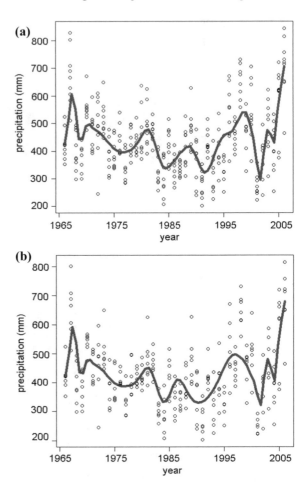

In Fig. 6 we present the wavelets model obtained for the data grouped as in *B*, for the annual regional precipitation (a) without Sulina and (b) Group I.

We remark the existence of a small difference between the shapes of the two curves, which appears in the period 1996–1998.

Analogous models have been built for the monthly regional precipitation series [2, 3]. MSE has been utilized for comparing their goodness of fit (Table 5).

When applying Method B, the mean standard errors decreases in all situations with a percentage between 18.2 % (for Entire) and 21.3 % (for Group I). Its values are between 37.27 and 39.81 when using A, and between 30.49 and 31.11, when using B [3].

Since some of the modeling procedures are non-parametric, their significance is estimated function of the *p*-values. Higher the *p*-value is, better the model is. The *p*-values corresponding to some of these models are provided in [3]. In all the study cases, the *p*-values corresponding to the models built after grouping as in *B* are higher than the others.

Table 5 MSE—monthly data (mm)

	Splines		Wavelets	
Series	A	B	A	B
Entire	26.13	24.56	37.27	30.49
Without Sulina	26.34	24.42	37.78	31.08
Group I	25.89	19.39	38.01	29.91
Group II	26.54	24.45	39.51	31.11

To have a general image of the evolution of the regional precipitation, we recommend the method *B*, because it provides a smoother curve that captures the essential variation of regional precipitation.

On the other hand, the Principal Component Analysis applied to the annual and monthly precipitation data points out that only two components are enough to describe the precipitation evolution [3]. Therefore, taking into account the small intervals of variation of the MSEs in the wavelets models, evaluation of regional precipitation evolution can be done using any group of data, without significant loss of accuracy, even if the best results are obtained with Group I. Moreover, if some data are not available for one station and the estimation of the precipitation evolution at regional level is known, the missing data can be replaced by those given by the regional model.

2.2 Models for the Regional Monthly Maximum Annual Precipitation

The modeling has been done for the monthly maximum annual precipitation series registered at the meteorological stations from Dobrogea in the period 1965–2005. The group formed by the main series is denoted in the following by Gr.AM (Fig. 7), and that composed by the secondary ones is denoted by Gr.BM (Fig. 8).

The Levene test applied to the series from Gr.AM confirmed the homoskedasticity hypothesis.

The randomness of the data series has been checked using the Wald–Wolfowitz test, at the significance level of 0.05.

Table 6 contains the results of Wald–Wolfowitz test for the series from Gr.AM. Since all the *p*-values (*p*-val) are greater than 0.05, the null hypothesis can not be rejected.

For the regional monthly maximum annual series formed by all the series from Gr.AM, but Sulina, the value of the Wald–Wolfowitz statistics is -1.432 and the corresponding *p*-value is $0.152 > 0.05$. Therefore, we can not reject the hypothesis that this series is random.

Table 7 contains the results of the same test for the secondary series. At the significance level of 0.05, the randomness hypothesis is rejected for three series (Lipnita, Negru Voda, and Pestera).

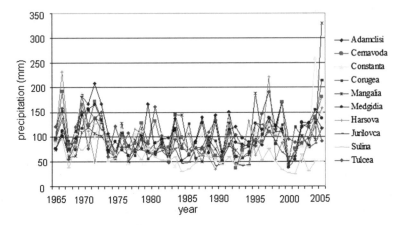

Fig. 7 Monthly maximum annual precipitation series record—Gr.AM

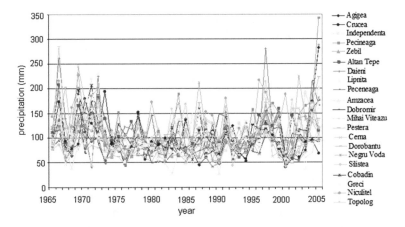

Fig. 8 Monthly maximum annual precipitation series record—Gr.BM

As expected, the normality hypothesis was rejected for all series. The hypothesis that the series are independent of each other can be sustained.

Even if ANOVA is robust to the violation of normality hypothesis, for testing the homogeneity in mean for the series in different groups, we also used the Kruskal-Wallis test, after ANOVA.

In Table 8 we find the results of ANOVA applied for Gr.AM and for Gr.AM without Sulina. It results that the equality of the populations' means can be assumed only in the case (b).

After performing the Tukey and Scheffé tests at the significance level of 0.05 we can not reject the hypothesis of the means' equality for all series from Gr.AM but Sulina, because the *p*-values are respectively 0.058 and 0.871 (Table 9).

The Kruskal–Wallis test applied for Gr.AM and Gr.AM without Sulina confirms the previous conclusions.

Table 6 Results of Wald–Wolfowitz tests for the series from Gr.AM

	Adamclisi	Cernavoda	Constanta	Corugea	Harsova
Test val. (a)	106.71	101.67	94.50	96.26	90.93
Case < test val.	23	25	22	22	26
Case >=test val	18	16	19	19	15
Case nos	41	41	41	41	41
Run nos.	18	18	18	21	17
Z	−0.87	−0.67	−0.92	0.00	−0.86
p-val. (bilat)	0.39	0.50	0.36	1.00	0.39
	Jurilovca	Medgidia	Mangalia	Tulcea	Sulina
Test val. (a)	85.41	101.18	95.67	104.09	60.57
Case < test val.	22	22	24	24	26
Case >=test val	19	19	17	17	15
Case nos	41	41	41	41	41
Run nos.	17	18	16	20	16
Z	−1.24	−0.92	−1.44	−0.13	−1.20
p-val. (bilat)	0.22	0.36	0.15	0.896	0.23

We did the same study for Gr.BM. At the significance level of 0.05, ANOVA rejected the null hypothesis of means' equality.

After applying the Tukey test, the following homogenous groups have been detected:

- Gr.BM.I formed by the series from Gr.BM but Negru Voda, Negureni, and Niculitel,
- Gr.BM.II that is Gr.BM without Casian, Negureni, and Niculitel,
- Gr.BM.III, which is Gr.BM without Casian, Casimcea, Corbu, Lumina, Mihai Viteazu, Negureni, Saraiu, Satu Nou,
- Negureni.

Performing the Scheffé test, the following homogenous groups have been detected:

- Gr.BM.IV, which is formed by the sets of Gr.BM but Negureni,
- Gr.BM.V that contains the series: Albesti, Altan Tepe, Biruinta, Daeni, Hamcearca, Independența, Lipnita, Negru Voda, Negureni, Niculitel, Pantelimon, Pecineaga, Topolog.

Performed on the groups Gr.BM.I–Gr.BM.V, the Kruskal–Wallis test did not reject the hypothesis of means' equality.

As in the previous section, we employed the wavelets technique for modeling the regional monthly annual precipitation, based on the detected groups. We present here the charts of the regional precipitation evolution using Gr.AM but Sulina (Fig. 9), Gr.BM.I–Gr.BM.V (Figs. 10, 11 and 12) with *Method A*.

Table 7 Results of the Wald-Wolfowitz test for the secondary series

	Agigea	Altan Tepe	Amzacea	Cerna	Cobadin	Crucea	Daieni
Test val. (a)	95.99	108.10	93.77	94.71	98.88	94.46	113.32
Case < test val.	26	23	24	23	24	28	26
Case ≥ test val	15	18	17	18	17	13	15
Case nos.	41	41	41	41	41	41	41
Run nos.	18	15	18	23	19	19	16
Z	0.521	−1.829	−0.783	0.419	−9.456	0.000	−1.204
p-val. (bilat)	0.603	0.067	0.433	0.675	0.647	1.000	0.229

	Dobromir	Dorobantu	Greci	Independenta	Lipnita	Mihai Viteazu	Negru Voda
Test val. (a)	95.28	97.33	93.61	105.19	104.56	84.58	115.28
Case < test val.	24	27	25	21	25	23	25
Case ≥ test val	17	14	16	20	16	18	16
Case nos.	41	41	41	41	41	41	41
Run nos.	17	22	20	20	13	19	12
Z	−1.109	0.727	−0.004	−0.313	−2.333	−0.545	−2.666
p-val. (bilat)	0.267	0.467	0.997	0.755	0.020	0.586	0.008

	Niculitel	Pecineaga	Peceneaga	Pestera	Silistea	Topolog	Zebil
Test val. (a)	127.25	107.85	96.12	96.79	95.18	110.14	96.92
Case < test val.	26	24	23	24	22	24	26
Case ≥ test val	15	17	18	17	19	17	15
Case nos.	41	41	41	41	41	41	41
Run nos.	18	14	25	11	22	23	15
Z	−0.521	−2.088	1.062	−3.066	0.035	0.521	−1.545
p-val. (bilat)	0.603	0.037	0.288	0.002	0.972	0.602	0.122

Table 8 ANOVA for (a) Gr.AM, (b) Gr.AM without Sulina

(a)	SS	df	MS	F	p-val.
Between groups	64,933.93	9	7214.881	5.105	0.000
Within groups	565,301.69	400	1413.254		
Total	630,235.62	409			
(b)	SS	df	SS	F	p-val.
Between groups	14,846.86	8	1855.85	1.232	0.279
Within groups	542,131.93	360	1505.92		
Total	556,978.79	368			

Table 9 Results of the Tukey and Scheffé tests

(a) Homogenous groups in Gr.AM

Series	Tukey-groups		Scheffé-groups	
	1	2	1	2
Sulina	60.54		60.54	
Jurilovca	85.41	85.41	85.41	85.41
Hârsova		90.93	90.93	90.93
Constanta		94.49	94.49	94.49
Mangalia		95.66		95.66
Corugea		96.26		96.26
Medgidia		101.18		101.18
Cernavoda		101.67		101.67
Tulcea		104.08		104.08
Adamclisi		106.70		106.70
p-val.	0.085		0.057	0.681

(b) Homogenous groups in Gr.AM without Sulina

	Tukey HSD	Scheffé
Jurilovca	85.41	85.41
Hârsova	90.931	90.93
Constanta	94.49	94.49
Mangalia	95.66	95.66
Corugea	96.26	96.26
Medgidia	101.18	101.18
Cernavoda	101.67	101.67
Tulcea	104.08	104.08
Adamclisi	106.70	106.70
p-val.	0.243	0.628

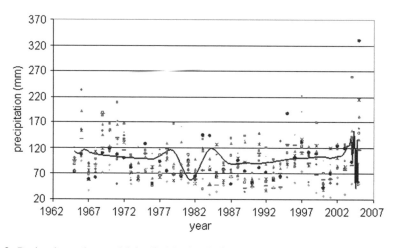

Fig. 9 Regional wavelets model for Gr.AM, but Sulina

Fig. 10 Regional wavelets
model respectively for:
a Gr.BM.I, **b** Gr.BM.II

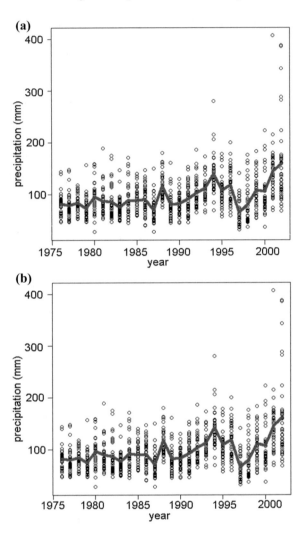

Due to the constraint imposed by the wavelets method concerning the number of values that can be used (which must be a power of 2), the modeling has been done for the most recent period taking into account the corresponding number of months.

Analysis of Figs. 10 and 11 shows that at the beginning of the period investigated there were small variations of the extreme values. We notice a period of high oscillations of the extreme values, after 1987. All these variations are well captured by models.

Fig. 11 Regional wavelets model respectively for: **a** Gr.BM.III and **b** Gr.BM.IV

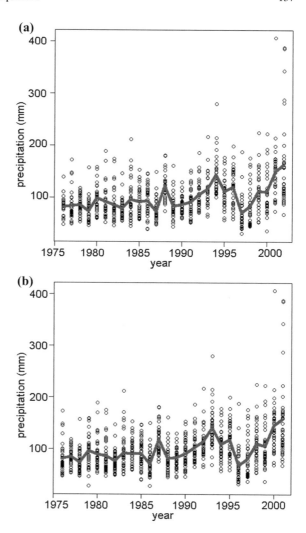

The models presented in Figs. 10, 11 and 12 do not differ significantly viewpoint of the mean standard errors (on the common period). Therefore, each of them can be used to describe the evolution of maximum monthly annual precipitation without a significant loss of information if some series are not available, or there are gaps in information.

Fig. 12 Regional model built
with the data from Gr.BM.V

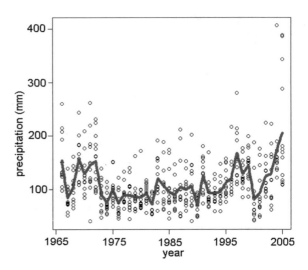

3   Generation of Annual Precipitation Field at Regional Scale

3.1   Overview of Some Theoretical Approaches

The elaboration of models for the generation of precipitation fields at local and
regional scale has become an important research subject last period, due especially
to the registered climatic change.

The results of the generation of precipitation fields at each main
hydro-meteorological station from Dobrogea have been discussed in Chap. 3. Here
we focus on two different aspects. The first one concerns the simultaneous estimation of AR parameters in the models for the annual precipitation (Chap. 3). The
second one treats the monthly precipitation series generation, applied to Dobrogea
region (for the period Jan 1965–Dec 2005).

One of the methods usually employed for generation the precipitation series at
regional scale is based on the model [36]:

$$X_t = AX_{t-1} + Be_t, \tag{1}$$

where X_t is a quadratic matrix that contains the standardized precipitations record
for the year t, e_t is a random vector of errors with zero mean and a unit standard

deviation, A and B are quadratic matrices of the same order, with constant coefficients which contain the autocorrelations and the cross-correlations.

If the data is skew, it can be normalized by means a log-normal transformation whose parameters are obtained by using the equations proposed by Matalas [22]. Also, the Eq. (1) can be simplified, by ignoring the cross-correlations of the first order [12].

The model (1) was criticized for its inadequacy of representing long dry periods or episodes with high rainfalls.

A new methodology for the stochastic simulation and forecasting of hydrological series has been introduced by Koutsoyiannis [16, 17, 19]. It uses a generalized autocorrelation function, implemented in a symmetric moving average scheme in which the parameters' number and the type of the autocovariance function are decided by the user [19].

Nonparametric models have been also proposed last period [10] due to their advantage that they do not rely on the distribution of the hydrological data nor on restrictive hypotheses concerning the dependence structure [30]. Examples of such models are: moving block bootstrap (MBB), k-nearest neighbor bootstrap (k-NN) [24, 25], methods based on kernels [25, 33], HMBB (hybrid moving block bootstrap) [30, 31], multi-modular [26].

Frost [10] introduced a Bayesian method for the calibration and evaluation of hydrological data at annual scale, at many sites, that uses HMM and AR(1) models. They are extended by introducing a Box–Cox transformation and a spatial correlation function. The models' performances are assessed by using the Bayesian selection method that evaluates the models' probabilities.

Another approach to the problem appears in [11], where the spatial interpolation of the parameters of AR(1) model is done using the *thin plates splines*.

Here, we use *lowess* for performing the spatial interpolation of the parameters of AR(1) models of annual precipitation data, presented in Chap. 3.

The terms *loess* [6] and *lowess* are the abbreviation of the terms *locally estimated scatterplot smoothing* and *locally weighted scatterplot smoothing* and designate nonparametric regression procedures. For fitting the data, both methods use locally weighted linear regression.

Loess estimates a regression surface using a multivariate procedure of fitting that build a smoothing local function that depends on the independent variables. For estimating a value at a given point, only the neighbor data is employed, analogous to the moving average procedure for the time series. The local fit permits the estimation of a larger class of surfaces than using the classic parametrical functions, as, for examples, the polynomials of different degrees [7].

The result of the application of the loess procedure is a smooth curve that locally minimizes the prediction errors or the residual variance.

The basic idea of the *lowess* algorithm [6] is to perform a local polynomial fit by means of the least squares method, followed by the use of the k-nearest neighbours method and by another robust method to obtain the final data estimation. More precisely, firstly one have to perform a local polynomial fitting in the neighbourhood of each point x, to estimate the vector of parameters $\beta \in R^{p+1}$, which minimizes the expression

$$n^{-1} \sum_{i=1}^{n} W_{ki}(z) \left(x_i - \sum_{j=0}^{p} \beta_j z^j \right)^2,   \tag{2}$$

where x_i are the values of the endogenous variable and $W_{ki}(z)$ are the weight associated with the neighbour values.

In the second phase, the residual, $\hat{\varepsilon}_i$, and the scale parameter $\hat{\sigma} =$ median $(\hat{\varepsilon}_i)$ are estimated. Then, the robustness weights are determined, by:

$$\delta_i = K(\hat{\varepsilon}_i/6\hat{\sigma}),$$

where:

$$K(u) = \begin{cases} (1-u)^3, & |u| \leq 1 \\ 0, & |u| > 1 \end{cases}.   \tag{3}$$

The final step is building a polynomial regression as at the first step, with the weights $\delta_i W_{ki}(z)$.

Cleveland suggested that a value $p = 1$ could assure the equilibrium between the computation speed and the flexibility in the reproduction of data patterns.

The smoothing parameter is computed by estimating the standard deviation or by the cross-validation procedure [27]. These procedures are implemented in R, in **stats** package [13], in the functions 'loess()' and 'lowess()'. The function 'lowess()' has initially implemented and uses locally-weighted polynomial regression. It contains an argument that permits to set up the number of values used for fitting the others. In 'loess()' function, a new formula is used: the fit at a given point is done using as weights the distances from the points that intervene in the computation to it.

3.2   Application to Annual Precipitation in Dobrogea Region

To estimate the precipitation evolution at all the stations in Dobrogea, we used the lowess method, to interpolate the parameters' in AR(1) models for each annual series.

The setting used are:

- The number of the neighbour stations for estimation: 50 or 100 %;
- The kernel, K, defined in (3),
- Methods of bandwidth selection: minimization of the mean standard error.

Table 10 Errors in lowess model

Series	50 % of neighbours				All the neighbours			
	Local polynomial's degree				Local polynomial's degree			
	0	1	2	3	0	1	2	3
Adamclisi	−0.078	−0.154	−0.154	−0.154	0.016	−0.023	−0.112	−0.112
Cernavodă	0.031	0.109	0.109	0.109	0.072	0.143	0.047	0.048
Constanţa	−0.044	−0.031	−0.031	−0.031	−0.005	0.000	0.018	0.038
Corugea	−0.106	−0.329	−0.329	−0.329	−0.094	−0.111	0.248	0.248
Harşova	−0.074	−1.107	−1.107	−1.107	−0.042	−0.134	−0.081	−0.081
Jurilovca	0.129	0.075	0.075	0.075	0.203	0.204	0.117	0.120
Mangalia	−0.395	0.459	0.459	0.459	−0.348	−0.448	−0.334	−0.380
Medgidia	0.035	0.101	0.101	0.101	0.031	−0.094	−0.030	0.032
Sulina	0.130	−0.076	−0.076	−0.076	0.146	−0.039	−0.031	−0.172
Tulcea	−0.051	−0.144	−0.144	−0.144	−0.001	0.220	0.256	0.213
MAE	0.107	0.259	0.259	0.259	0.0957	0.1417	0.1275	0.1445
MSE	0.147	0.402	0.402	0.402	0.141	0.187	0.167	0.179

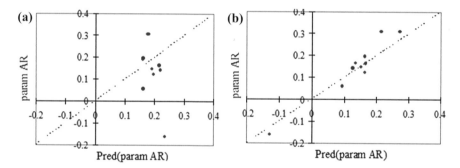

Fig. 13 Comparison of the parameters in AR models (param AR) and the estimated ones [Pred (param)] in lowess approach with: **a** 50 % neighbours or **b** 100 % neighbours used

The results presented in Table 10 prove that the best model has been obtained by using a local constant smoothing and all the neighbours.

The dotted line in Fig. 13 represents the first bisectrix of axes. Better the estimation is, closer to the bisectrix the dots are.

Remark the good data distribution, especially in the case (b). Therefore, the estimated values of AR coefficient could be used to generate the precipitation fields.

Better results have been determined using the Gaussian kernel.

In the following we present the code in R for obtaining the estimation.

```
data<-read.csv("D:\\Lucrari_2.12.14\\2015_Carte\\loess_nou.csv",sep=",",header=TRUE)
lat<-data[,2]
lon<-data[,3]
alt<-data[,4]
AR<-data[,5]
AR.loess<- loess(AR ~ lon + lat, span=1)
summary(AR.loess)
```

Call:
loess(formula = AR ~ lon + lat, span = 1)

Number of Observations: 10
Equivalent Number of Parameters: 6.2
Residual Standard Error: 0.1295
Trace of smoother matrix: 7.3

Control settings:
normalize: TRUE
span   : 1
degree : 2
family : gaussian
surface : interpolate   cell = 0.2

cor(AR, AR.loess$fitted)^2

[1] 0.9002

References

1. Bărbulescu, A., Deguenon, J., Modeling the annual precipitation evolution in the region of Dobrudja. Int. J. Math. Models Methods Appl. Sci. **6**(5), 617–624 (2012)
2. Bărbulescu, A., Deguenon, J.: Nonparametric methods for fitting the precipitation variability applied to Dobrudja region. Int. J. Math. Models Methods Appl. Sci. **6**(4), 608–615 (2012)
3. Bărbulescu, A., Deguenon, J.: Models for trend of precipitation in Dobrudja. Env. Eng. Manage. J. **13**(4), 873–880 (2014)
4. Bărbulescu, A., Deguenon, J., Teodorescu, D.C.: Study on water resources in the Black Sea region. Nova Publishers, USA (2011)
5. Basivi, R., Zawadzki, I., Fabry, F.: Predictability of precipitation from continental radar images. Part V: Growth Decay J. Atmos. Sci. **69**, 3336–3349 (2012)
6. Cleveland, W.S.: Robust locally weighted regression and smoothing scatterplots. J. Am. Stat. Assoc. **74**, 829–836 (1979)
7. Cleveland, W.S., Devlin, S.J.: Locally weighted regression: an approach to regression analysis by local fitting. J. Am. Stat. Assoc. **83**(403), 596–610 (1988)
8. Duan, M., Ma, J., Wang, P.: Preliminary comparison of the CMA ECMWF, NCEP, and JMA Ensemble Prediction Systems. Acta Meteorologica Sinica **26**(1), 26–40 (2012)
9. Fox, J.: Multiple and generalized nonparametric regression. Sage, Thousand Oaks CA (2000)

10. Frost, A.J., Thyer, M.A., Srikanthan, R., Kuczera, G.: A general Bayesian framework for calibrating and evaluating stochastic models of annual multi-site hydrological data. J. Hydrol. **340**, 129–148 (2007)
11. Hancock, P. A., Hutchinson, M. F.: Thin plate smoothing spline interpolation of parameters of the AR(1) annual rainfall model across the australian continent, working document 02/7. Cooperative Research Centre for Catchment Hydrology (2002)
12. Hipel, K. W.: Stochastic research in multivariate analysis. In: Fourth International Hydrology Symposium, pp. 2–50. Fort Collins, Colorado, USA
13. https://stat.ethz.ch/R-manual/R-patched/library/stats/html/loess.html
14. Jo, S., Lim, Y., Lee, J., Kang, H.S., Oh, H.S.: Bayesian regression model for seasonal forecast of precipitation over Korea. Asia-Pacific J. Atmos. Sci. **48**(3), 205–212 (2012)
15. Katz, R.W.: Statistical procedures for making inferences about precipitation changes simulated by an atmospheric general circulation model. J. Atmos. Sci. **40**, 2193–2201 (1983)
16. Koutsoyiannis, D.: A generalized mathematical framework for stochastic simulation and forecast of hydrologic time series. Water Resour. Res. **36**(6), 1519–1533 (2000)
17. Koutsoyiannis, D., Manetas, A.: Simple disaggregation by accurate adjusting procedures. Water Resour. Res. **32**, 2105–2117 (1996)
18. Kruskal, W.H., Wallis, W.A.: Use of ranks in one-criterion variance analysis. J. Am. Stat. Assoc. **47**(260), 583–621 (1952)
19. Langousis, A., Koutsoyiannis, D.: A stochastic methodology for generation of seasonal time series reproducing overyear scaling behavior. J. Hydrol. **322**, 138–154 (2006)
20. Lorenz, E.N.: Deterministic non-periodic flow. J. Atmos. Sci. **20**, 130–141 (1963)
21. Luo, L., Wood, E.F.: Assessing the idealized predictability of precipitation and temperature in the NCEP Climate Forecast System. Geophys. Res. Lett. **33**, L04708 (2006)
22. Matalas, N.C.: Mathematical assessment of synthetic hydrology. Water Resour. Res. **4**(3), 937–945 (1967)
23. Popescu-Bodorin, N., Bărbulescu, A.: Crisp and fuzzy history-based predictability of regional monthly precipitation data (under review)
24. Prairie, J., Rajagopalan, B., Lall, U., Fulp, T.: A stochastic nonparametric technique for space–time disaggregation of streamflows. Water Resour. Res. **43**, W03432 (2007). doi:10.1029/2005WR004721
25. Rajagopalan, B., Lall, U.: A k-nearest—neighbor simulator for daily precipitation and other weather variables. Water Resour. Res. **35**(10), 3089–3101 (1999)
26. Serinaldi, F., Kilsby, C.G.: A modular class of multisite monthly rainfall generators for water resource management and impact studies. J. Hydrol. **464–465**, 528–540 (2012)
27. Silverman, B.W.: Some aspects of the spline smoothing approach to non-parametric regression curve fitting. J. R. Stat. Soc. Series B Stat. Methodol. **47**(1), 1–52 (1985)
28. Skaug, H.J., Tjøstheim, D.: Nonparametric tests of serial independence. In: Subba Rao, T. (ed.) Developments in time series analysis: The Priestley birthday volume, pp. 207–229. Chapman & Hall, London (1993)
29. Sloughter, J.M., Raftery, A.E., Gneiting, T., Fraley, C.: Probabilistic quantitative precipitation forecasting using Bayesian model averaging. Mon. Weather Rev. **135**, 3209–3220 (2007)
30. Srinivas, V.V., Srinivasan, K.: Hybrid moving block bootstrap for stochastic simulation of multi-site multi-season streamflows. J. Hydrol. **302**, 307–330 (2005)
31. Srinivas, V.V., Srinivasan, K.: Hybrid matched-block bootstrap for stochastic simulation of multiseason streamflows. J. Hydrol. **329**, 1–15 (2006)
32. Szekely, G.J., Rizzo, M.L., Bakirov, N.K.: Measuring and testing dependence by correlation of distances. Ann. Stat. **35**(6), 2769–2794 (2007)
33. Tarboton, D.G., Sharma, A., Lall, U.: Disaggregation procedures for stochastic hydrology based on nonparametric density estimation. Water Resour. Res. **34**, 107–119 (1998)
34. Urs, G., Zawadzki, I., Turner, B.: Predictability of precipitation from continental radar images. Part IV: Limits to prediction. J. Atmos. Sci. **63**, 2092–2108 (2006)

35. Wu, L., Seo, D.J., Demargne, J., Brown, J.D., Conga, S., Schaake, J.: Generation of ensemble precipitation forecast from single-valued quantitative precipitation forecast for hydrologic ensemble prediction. J. Hydrol. **399**, 281–298 (2011)
36. Young, G.K., Pisano, W.C.: Operational hydrology using residuals. J. Hydraul. Div. ASCE **94** (HY4), 909–923 (1968)

Chapter 5
Analysis and Models for Surface Water Quality

In this chapter we discussed about the stationarity of the series of hydrochemical properties of the Techirghiol Lake (situated in Dobrogea region, Romania) and we summarize the results concerning the evolution of temperature of Techirghiol Lake's water and the relationship between it and the atmospheric temperature.

The quality of the surface waters is of big importance for the human health and economy. The anthropic activities and the atmospheric pollution have negative effects on the water quality in the urban zones. The ecosystems of the lakes from Constanta County, Romania, are also affected by these factors.

Knowledge on the pollutants' effects on the surface and underground water and their transportation is the first step in the management of water quality. Precipitation is one of the main means of pollutants' movement from the air to soil and surface water. As an effect of the accumulation of different kinds of pollutants, many urban lakes suffer an eutrophication process. One of these cases is that of Tabacarie Lake from Constanta city (Fig. 1).

To establish the influence of rainwater's composition on that of the urban lakes, we conducted an experiment on Tabacarie Lake. Data were recorded and processed for the period January 2000–June 2008. They consisted of pH and chemical oxygen demand for the lake's water, and pH, the concentration of chlorine, ammonium, sulfate, the acidity/alkalinity and the conductivity of rainwater collected at the same location.

The study revealed the dependence of the rainwater's pH on the rainwater's alkalinity, of the rainwater's pH on NH_4^+ and SO_4^{2-} concentrations and of the pH of the lake's water on the rainwater's pH [2].

In this chapter we present some results of our study about the evolution of the quality of the water of Techirghiol Lake.

© Springer International Publishing Switzerland 2016
A. Bărbulescu, *Studies on Time Series Applications in Environmental Sciences*,
Intelligent Systems Reference Library 103, DOI 10.1007/978-3-319-30436-6_5

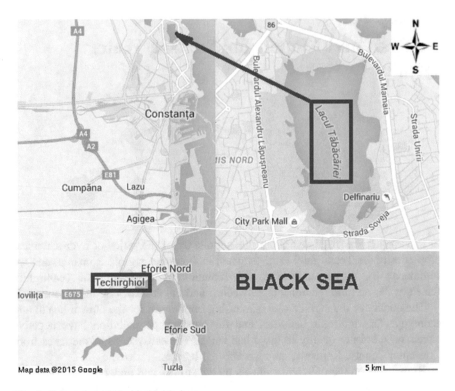

Fig. 1 Tabacarie and Techirghiol Lakes

1   On the Hydrochemical Properties of the Water of Techirghiol Lake

Techirghiol Lake (Fig. 1) is situated on the Black Sea Littoral, at approximately 15 km South from Constanta city. It is a para-littoral lake with a surface of 10.68 km^2, a length of 7500 m, the maximum depth of 12 m, and a hypersaline system. It has a curative importance, due to the existence of the sapropelic mud, unique in Romania, which is the result of the bacterial decomposition of Artemia salina filopodia and the Cladophora euryhaline algae. Also, it belongs to the European network of protected areas, Natura 2000. Due to its importance, its hydrochemical, morphometric and ecological features have been extensively studied [7, 9–11, 15, 17].

In [3] we analyzed some hydrochemical characteristics of this lake, recorded between 1991 and 2011 (Fig. 2), in the context of the existence of periods of water influx from precipitation or irrigation. They are the salinity, the dried residue, the dissolved oxygen (DO), the biochemical oxygen demand (BOD$_5$), the chemical oxygen demand by potassium permanganate method (COD-Mn), the total hardness and the concentration of $Ca^{2+}, Cl^-, HCO_3^-, Mg^{2+}$.

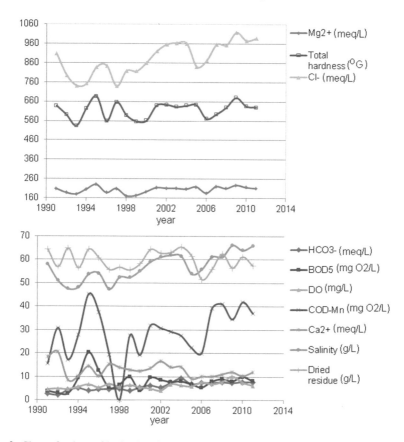

Fig. 2 Chart of values of hydrochemical characteristics of Techirghiol Lake

After the tests of normality, homoscedasticity, autocorrelation, breakpoints and outliers' detection, we applied the non-parametric Mann-Kendall test for testing the existence of a linear trend of the data series. For the series for which the hypothesis of the trend existence wasn't rejected, the linear trend has been determined, by Sen's non-parametric method [16]. The equations of these curves are [3]:

- For salinity (significant at 0.001):

$$y_{t\,(sal)} = 0.87t + 47.38,$$

- For Cl$^-$ (significant at 0.001):

$$y_{t(Cl)} = 11.56t + 758.09,$$

- For DO (significant at 0.01):

$$y_{t(\mathrm{DO})} = 0.13t + 4.71,$$

- For Mg^{2+} (significant at 0.05):

$$y_{t(\mathrm{Mg})} = 1.20t + 194.17,$$

- COD-Mn (significant at 0.1):

$$y_{t(\mathrm{COD-Mn})} = 0.55t + 24.60.$$

Maintaining the concentrations of the chemical elements in constant limits is an important issue for preserving the ecological conditions of Techirghiol Lake for producing the mud with therapeutical properties. Therefore, we address here the stationary of the series of concentrations of different elements.

Table 1 The p-values in ADF and KPSS tests for the studied series

Series	ADF (a)	KPSS (b)		Remark
		Level	Trend	
Salinity	0.2607	<0.01	>0.1	(a) Unit root (b) Nonstationarity in level Stationarity in trend
Cl^-	0.2343	<0.01	>0.1	(a) Unit root (b) Nonstationarity in level Stationarity in trend
HCO_3^-	0.7521	<0.01	>0.1	(a) Unit root (b) Nonstationarity in level Stationarity in trend
DO	0.6684	0.0208	>0.1	(a) Unit root (b) Nonstationarity in level Stationarity in trend
Mg^{2+}	0.2024	0.0884	>0.1	(a) Unit root (b) Stationarity in level Stationarity in trend (c) PP-test: p-val = 0.0165
COD-Mn	0.5476	>0.1	>0.1	(a) Unit root (b) Stationarity in level Stationarity in trend (c) PP-test: p-val = 0.035
Ca^{2+}	<0.01	>0.1	>0.1	Stationarity
BOD_5	0.0231	>0.1	>0.1	Stationarity
Dried residue	0.3686	>0.1	>0.1	(a) Unit root (b) Stationarity in level Stationarity in trend (c) PP-test: p-val = 0.0175
Total hardness	0.0214	>0.1	>0.1	Stationarity

We apply the ADF [8] test for checking the existence of a unit root, against the stationarity. Then, as a confirmatory test, we use the KPSS test [13] to verify the stationarity hypothesis. Phillips—Perron test [14] can also be used for this purpose. The truncation lag is considered to be 2.

Table 1 contains the p-values corresponding to these tests.

For the first four series, ADF could not reject the hypothesis of the existence of a unit root, and KPSS rejected the hypothesis of stationarity in level, but could not reject that of the trend stationarity.

For Ca^{2+}, BOD_5 and *total hardness*, the stationarity hypothesis could not be rejected. For Mg^{2+}, COD-Mn and *dried residue*, the results of the two tests are not concordant. Therefore, we apply the PP test [14] because its statistics takes into account the autocorrelation and heteroskedasticity. Its results are concordant with those of KPSS, so we can not reject the stationarity hypothesis of for Mg^{2+}, COD-Mn and *dried residue*.

In the last column of Table 1, we resume the findings of ADF (a), KPSS (b) and, where it is the case, of PP (c).

The first four series, for which a significant increasing trend has been determined, are non-stationary in level, but stationary in trend. Therefore, for presserving the water quality, the evolution of the salinity, Cl^-, HCO_3^- and DO must be permanently monitored and actions for limiting their variations must be taken.

2 On the Temperature of the Water of Techirghiol Lake

Studies concerning the temperature evolution in Dobrogea region are noticed the last period [4, 5, 12]. However, only two recent articles [1, 6] refer to Techirghiol city, where the lake with the same name is situated. In these studies we propose models for:

- the daily average air temperature in the period 1.01.1961–31.12.2009 at Techirghiol;
- the monthly average air temperature in the interval Jan 1961–Dec 2009 at Techirghiol;
- the daily average temperature of the water of Techirghiol Lake in 2010;
- the daily average temperature of the water of Techirghiol Lake function of daily atmospheric temperature for the period 1.01.2010–31.12.2010;

The first two models are built for the detrended data, using GRNN, with the proportion of 90:10 of the training and the validation datasets. We don't discuss them here, being out the aim of this chapter.

The model for the trend of the average daily temperature of the water (y_t) is:

$$\hat{x}_t = 12.1946 + 13.1506 \cos(0.0165t - 3.3654).$$

with a standard error of 2.43 [1].

Two models have been built for describing the dependence of water temperature (y_t) on that of the air temperature (x_t):

- a linear one, with the equation:

$$y_t = 2.445x_t + 0.923 + \varepsilon_t, t = \overline{1,365},$$

where (ε_t) is the residual, and the determination coefficient is $R^2 = 0.904$ [1];
- a GRNN model, with the following performances on the training and validation datasets [6]:

  - the correlation between the actual and predicted values (%), respectively 96.29 and 96.05;
  - the root mean square errors, respectively 2.59 and 2.67;
  - the mean absolute errors, respectively 2.01 and 2.08.

They can be successfully used for forecasting the dependence of the temperature of water of Techirghiol Lake on air temperature.

For detail, the reader may see [1, 6].

References

1. Bărbulescu, A.: Modeling temperature evolution. Case study, Rom. Rep. Phys. **68**(1) (2016) (to appear)
2. Bărbulescu, A.: The analysis of correlation of some ions concentration in rainwater in an urban area. Int. J. Math. Models Methods Appl. Sci. **4**(2), 105–112 (2010)
3. Bărbulescu, A., Barbeş, L.: Assessment of surface water quality of Techirghiol Lake using statistical analysis. Rev. Chim. **64**(8), 868–874 (2013)
4. Bărbulescu, A., Băutu, E.: ARIMA and GEP models for climate variation. Int. J. Math. Comput. **3**(J09), 1–7 (2009)
5. Bărbulescu, A., Băutu, E.: Mathematical models of climate evolution in Dobrudja. Theor. Appl. Climatol. **100**(1–2), 29–44 (2010)
6. Bărbulescu, A., Maftei, C.: Modeling the climate in the area of Techirghiol Lake (Romania). Rom. J. Phys. **60**(7–8), 1163–1170 (2015)
7. Breier, A.: The Lakes from the Romanian Black Sea Coast. R.P.R. Academy Press, Bucharest (1976). (in Romanian)
8. Fuller, W.A.: Introduction to Statistical Time Series. Wiley, New York (1996)
9. Gastescu, P., Bretcan, P.: The Lakes In Romania Regional Limnology. R.S.R Academy Press, Bucharest (1971). (in Romanian)
10. Gastescu, P.: The lakes from the P. R. of Romania. Genesis and Hydrological Regime. R.P.R Academy Press, Bucharest (1963) (in Romanian)
11. Gheorghievici, L.M., Pompei, I., Gheorghievici, G., Tănase, I.: The influence of abiotic factors on suppliers of organic matter in the peloidogenesis process from Lake Techirghiol. Rom. Aquac. Aquarium, Conserv. Legislation **5**(2), 69–78 (2012)
12. Maftei, C., Bărbulescu, A.: Statistical analysis of climate evolution in Dobrudja region. In: Proceedings of World Congress on Engineering, Imperial College of London, England, 02–04 Jul 2008. Book series: Lecture Notes in Engineering and Computer sciences, vol. II, pp. 1082–1087 (2008)

13. Pfaff, B.: Analysis of Integrated and Cointegrated Time Series with R, 2nd edn. Springer, Heidelberg (2008)
14. Phillips, P.C.B., Perron, P.: Testing for a unit root in time series regression. Biometrika **75**(2), 335–346 (1998)
15. Romanescu, G: The ecological characteristics of the Romanian littoral lakes-the sector Midia Cape-Vama Veche. Lakes Reservoirs and Ponds **1–2**, 49–60 (2008)
16. Sen, P.K.: Estimates of the regression coefficient based on Kendall's Tau. J. Am. Stat. Assoc. **63**, 1379–1389 (1968)
17. Telteu, C.-E., Zaharia, L.: Morphometrical and dynamical features of the South Dobrogea lakes, Romania. Procedia Environ. Sci. **14**, 164–176 (2012)

Chapter 6
Models for Pollutants Dissipation

In this chapter, we present two types of models for the atmospheric pollutants' dissipation in the northern part of the Romanian Littoral. They are respectively:

- GRNN models for the evolution of NO_x, SO_x concentrations;
- GRNN models for the evolution of VOCs concentrations, function of atmospheric factors;
- Linear models for the dependence of the pollutants' concentrations on the urban zones on the pollutants' concentrations at the emissions sites;
- Linear model for the regional dispersion of H_2S–SO_2.

Atmospheric and soil pollution are ones of the most important issues that influence the human health and life. They are the results of industrial production, traffic and sewage sludge disposal. Pollutants' circulation also affects the plants that can accumulate different toxic elements on foliage and/or bark [1, 11, 12, 17].

Researchers use different plants as biomonitors of pollution in the urban area, as acacia, *Basella Alba*, lichens, mushrooms [9, 14, 15, 16] and so forth. One of them is *Populus nigra* sp., suitable as a bioindicator of heavy metals in Europe [10].

In [2, 3] we presented the results of our research on the metal trace accumulation in the bark and leaves of *Populus nigra L*. Moreover, we did comparisons with their accumulation in soil, in an area from the northern part of the Black Sea Littoral. We found a high correlation between the accumulation of heavy metals in leaves and soil, so the leaves of *Populus nigra L* act as accumulators.

The atmosphere is the largest space where pollutants propagate, their dissipation being closely related to the climate conditions, as wind, temperature, precipitation and air humidity. The major air pollutants are sulfur oxides (SO_x), nitrogen oxides (NO_x), carbon oxides (CO, CO_2), volatile organic compounds (VOCs) and particulate matter (PM_{10}, $PM_{2.5}$).

© Springer International Publishing Switzerland 2016
A. Bărbulescu, *Studies on Time Series Applications in Environmental Sciences*,
Intelligent Systems Reference Library 103, DOI 10.1007/978-3-319-30436-6_6

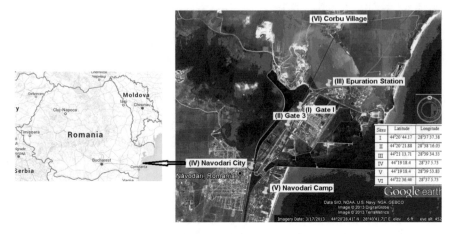

Fig. 1 Navodari area and the study sites

Many countries established regulations concerning the air quality monitoring and control, with the aim of maintaining a clean environment; Romania is not an exception. Last period, preservation of air quality on the Romanian Littoral has become a priority, due to the existence of a high population agglomeration, the existence of industrial parks as well as of the touristic resorts.

Our study, developed from 2006, addresses the actual issue of monitoring the pollutants' concentration on the Northern part of the Romanian Littoral, modelling and forecasting their evolution. For this aim, six monitoring sites (I–VI) have been established (Fig. 1). Data acquisition has been done as explained in [4, 6].

In the following, we summarize our finding, grouping them based on the modeling techniques employed.

1 GRNN Models

The first group of models developed implied the use of GRNN. We chose this approach because the classical techniques didn't provide satisfactory results.

For SO_2 and NO_2 monthly record (Jan 2007–Jun 2008) we tested the existence of a linear trend of the data [4]. For SO_2 series, this hypothesis has been rejected. For NO_2 series, it has been rejected for all but the series from Sites V and VI.

The exponential function provided a better (and more realistic) alternative to the linear trend [5]. Therefore, for the detrended data, GRNN models have been built. The associated MSEs and MAEs are presented in Table 1, proving the good estimation of the record data.

Models for the evolution of VOCs' concentration function of the lag one concentration, the variation of temperature, the wind speed and the air humidity are presented in [7]. We choose these predictors because the concentration of a

Table 1 MSEs and MAEs in GRNN models for NO_2 and SO_2

	Site	I	II	III	IV	V	VI
NO_2	MSE	2.0e−07	7.21e−07	0	0.00017	3e−006	0.00001
	MAE	0.00011	0.00021	0	0.01116	0.00056	0.00240
SO_2	MSE	4.8e−06	0.0000384	1.1e−06	9.74e−12	0.000076	7.39e−08
	MAE	0.000519	0.0024148	0.0005858	8.96e-07	0.002053	0.0000641

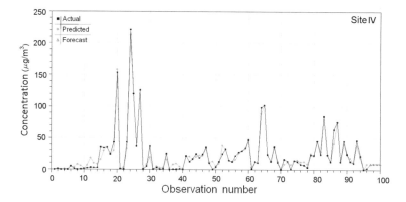

Fig. 2 GRNN model for VOCs

pollutant at a moment depends on its concentration previously existent in the atmosphere, as well as on the atmospheric conditions that influence its transport.

For modeling purposes, the series was divided in two, in the proportion 90:10, for training and validation. For training the network, we used the conjugate gradient descent algorithm [13].

Figure 2 contains the values recorded at Site IV, the estimated ones and the forecast for the next three months. In Fig. 3 we present the estimated concentrations with respect to the recorded ones at Sites I-III. The dots representing the estimated values are very close to the first bisectrix of the axes, proving that the models fit very well the data.

In Table 2 we present the MSEs and MAEs in GRNN models for the evolution of VOCs' concentrations.

The best models are those for Sites I and II. The worst model is that for the series from Site III, due to the existence of two extreme values (that are at least five times higher than the other values), that can not be captured by the models.

Fig. 3 Predicted versus actual values in GRNN models for VOCs collected at Sites I–Sites III

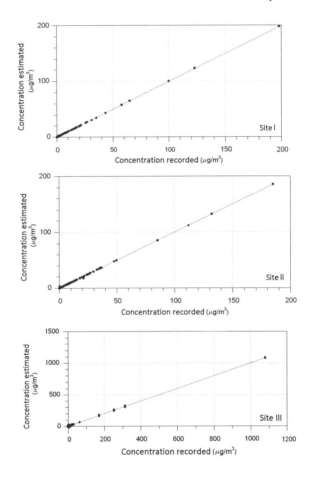

Table 2 MSE and MAE in GRNN models

	Site I	Site II	Site III	Site IV	Site V	Site VI
MSE	0.168	0.221	72.42	35.165	7.075	5.153
MAE	0.04	0.08	4.98	3.62	0.27	0.23

2   Linear Models

Knowing that the pollutants are mainly produced in the industrial zones (Sites I-III) we were interested in analysing their dissipation. Therefore, we proposed linear models for the propagation of CO, H_2S–SO_2, NO_x, and PM_{10} series registered in the period Nov 2007–Jul 2009 from the industrial to the urban zones.

Significant correlations were not detected between the H_2S–SO_2 concentration at the production and the reception points as well as for CO produced at Site I and NO_x produced at the Site III.

Linear models for the dependence between the pollutants' concentration at the production point and their concentration at the reception ones have been determined for:

- The dependence of CO concentrations at IV (V) on those at II,
- The dependence of NO_x concentrations at IV (V) on those at II and III,
- The dependence of PM_{10} concentrations at IV (V) on those at II and III.

For details, see [6].

Another approach was the determination of regional models for the dissipation of CO, NO_x and SO_2–H_2S, function of the atmospheric factors (wind speed, air temperature and humidity). So, we built a model:

$$Y_i = \alpha_i + \beta_1 v_i + \beta_2 H_i + \beta_3 T_i + \varepsilon_i, \tag{1}$$

where Y_i, v_i, T_i are respectively the pollutants' concentration, the wind speed and the temperature registered at the site i and ε_i is the random variable.

Eq. (1) is equivalent with:

$$y_{ij} = \alpha_i + \beta_1 v_{ij} + \beta_2 h_{ij} + \beta_3 t_{ij} + e_{ij}, \tag{2}$$

where $y_{ij}, v_{ij}, h_{ij}, e_{ij}$ are respectively the pollutants' concentration, the wind speed and the temperature, registered at the site i, at the moment j, and the corresponding residual.

For example, the exogenous variables in the model for H_2S–SO_2 were the air humidity and temperature and the logarithm of the wind speed. The t and F tests have been used to check the significance of each coefficient and of the model as a whole. Therefore, the parameters of the model are respectively:

$$\hat{\alpha}_1 = -7.824, \hat{\alpha}_2 = -7.207, \hat{\alpha}_3 = -7.280,$$

$$\hat{\alpha}_4 = -6.729, \hat{\alpha}_5 = -7.167, \hat{\alpha}_6 = -7.826,$$

$$\hat{\beta}_1 = 0.011, \hat{\beta}_2 = 0.0634, \hat{\beta}_3 = 0.6112$$

The residual in the model satisfy the normality, independence and homoscedasticity hypotheses.

Other models of regional of pollutants are presented in [8].

References

1. Ataabadi, M., Hoodaji, M., Najafi, P., Adib, F.: Evaluation of airborne heavy metal contamination by plants growing under industrial emissions. J. Environ. Eng. Manag. **9**(7), 903–908 (2010)

2. Barbeş, L., Bărbulescu, A., Rădulescu, C., Stihi, C., Chelărescu, E.D.: Determination of heavy metals in leaves and bark of Populus nigra L by atomic absorbtion spectrometry. Rom. Rep. Phys. **66**(3), 877–886 (2014)
3. Barbeş, L., Bărbulescu, A.: Monitoring and assessement of heavy metals in soil and leaves of Populus nigra L. J. Environ. Eng. Manage. http://omicron.ch.tuiasi.ro/EEMJ/accepted.htm
4. Bărbulescu, A., Barbeş, L.: Mathematical models for inorganic pollutants in Constanţa area, Romania. Rev. Chim. **64**(7), 747–753 (2013)
5. Bărbulescu, A., Barbes, L.: Models for pollutants evolution in an urban area. Math. Models Methods Appl. Sci. 151–156 (2012)
6. Bărbulescu, A., Barbeş, L.: Models for the pollutants correlation in the Romanian littoral. Rom. Rep. in Phys. **66**(4), 1189–1199 (2014)
7. Bărbulescu, A., Barbes, L.: Characterization of the concentrations of volatile organic compounds in the Romanian littoral using general regression neural networks: A case study. Anal. Lett. **40**(3), 387–399. doi:10.1080/00032719.2015.1027897
8. Bărbulescu, A., Barbes, L.: Modeling the pollutants dissipation. In: Proceedings of the 14th International Conference on Environmental Science and Technology Rhodes, Greece, 3–5 Sept 2015. http://cest.gnest.org/cest15proceedings/public_html/papers/cest2015_00435_poster_paper.pdf
9. Dulama, I., Popescu, I.V., Stihi, C., Radulescu, C., Cimpoca, G.V., Toma, L.G., Stirbescu, R., Nitescu, O.: Studies on accumulation of heavy metals in Acacia leaf by EDXRF. Rom. Rep. Phys. **64**(4), 1063–1071 (2012)
10. Kovács, M.: Biology Indicators in Environmental Protection. Ellis Horwood, New York (1992)
11. Markert, B., Breure, A.M., Zechmeister, H.G.: Bioindicators and Biomonitors: Principles, Concepts, and Applications. Elsevier, New York (2003)
12. Rossini Oliva, S., Fernandez Espinosa, A.J.: Monitoring of heavy metals in topsoils, atmospheric particles and plant leaves to identify possible contamination sources. Microchem. J. **86**(1), 131–139 (2007)
13. Shewchuk, J.R.: http://www.cs.cmu.edu/~quake-papers/painless-conjugate-gradient.pdf
14. State, G., Popescu, I.V., Gheboianu, A., Radulescu, C., Dulama, I., Bancuta, I., Stirbescu, R.: Identification of air pollution elements in lichens used as bioindicators by the XRF and AAS methods. Rom. J. Phys. **56**(1–2), 240–249 (2011)
15. Stihi, C., Popescu, I.V., Busuioc, G., Badica, T., Olariu, A., Dima, G.: Particle induced X-ray emission (PIXE) analysis of Basella Alba L leaves. J. Radioanal. Nucl. Chem. **246**(2), 445–447 (2000)
16. Stihi, C., Radulescu, C., Busuioc, G., Popescu, I.V., Gheboianu, A., Ene, A.: Studies on accumulation of heavy metals from substrate to edible wild mushrooms. Rom. J. Phys. **56**(1–2), 257–264 (2011)
17. Wolterbeek, B.: Biomonitoring of trace element air pollution: principles, possibilities and perspectives. Environ. Pollut. **120**(1), 11–21 (2002)

Chapter 7
Spatial Interpolation with Applications

Abstract Different spatial interpolation techniques are shortly discussed and applied on different precipitation series recorded in Dobrogea. We also discuss the interpolation of the parameters of AR models that describe the evolution of the annual precipitation series in the study region.

Different spatial interpolation techniques are shortly discussed and applied on different precipitation series recorded in Dobrogea. We also discuss the interpolation of the parameters of AR models that describe the evolution of the annual precipitation series in the study region.

A new method—MPPM—is also applied for the estimation of regional precipitation and its performances are compared with those of the Thiessen polygons method, IDW and ordinary kriging on Spring series. This method seems to be a good alternative for the Thiessen polygons methods. Its performances are also superior to those of the kriging and IDW on the study data, but more experiments must be carry on to confirm these findings.

1 Theoretical Considerations on the Spatial Interpolation Methods

Spatial interpolation methods are usually used for predicting the values of environmental variables in locations where the values are unknown, using their values in neighbor locations. Their specificity comes from incorporating information related to the geographic position of the sample data points.

Commonly used methods are classified in the categories [17, 18]: (1) deterministic methods, as: Nearest Neighbor, IDW, Splines, Classification, Regression, (2) geostatistical methods, as: Ordinary Kriging (KG), Universal Kriging, or (3) combined methods, as regression kriging, classification combined with other interpolation methods etc.

© Springer International Publishing Switzerland 2016   159
A. Bărbulescu, *Studies on Time Series Applications in Environmental Sciences*,
Intelligent Systems Reference Library 103, DOI 10.1007/978-3-319-30436-6_7

The spatial interpolation problem can be formulated as follows.

Consider a set of n observations, $z(s_1), z(s_2), \ldots, z(s_n)$, of a target variable Z, where $s_i = (x_i, y_i)$, $i = \overline{1, n}$ are the locations whose geographic coordinates are (x_i, y_i). If A is the study area and suppose that the observations samples are representative and consistent, determine the value of the target variable in a new location, s_0, function of $z(s_1), z(s_2), \ldots, z(s_n)$.

Therefore, the spatial interpolation model can be written as:

$$\hat{z}(s_0) = E\{Z | z(s_i), q_k(s_0), \gamma(h), \quad s_0 \in A\}$$

where: $\hat{z}(s_0)$ is the estimated value at the study point, E is the expected value of Z, $q_k(s_0)$ are deterministic predictors and $\gamma(h)$ is the covariance [15].

1.1 Mechanical Methods

1.1.1 The Thiessen Polygons Method [27]

The Thiessen polygons method relies on dividing the study region where the meteorological stations are situated, in polygons such as every point of the area is contained in a zone attached to the closest neighbor station. Building the Thiessen polygons is done in four steps:

- Draw on a map the region's boundaries and mark the stations' positions;
- Connect by segments the adjacent stations;
- Draw the mediators of the segments that connect the stations and connect them for determining the polygon corresponding to each station.

For computing the average precipitation in a study area, one has to perform the steps:

- Compute the surface of each polygon.
- Multiply the value of precipitation registered at a station by the area of the corresponding polygon;
- Sum the values previously obtained and divide the result by the surface of the region.

In Fig. 1 we present the partition of Dobrogea region in polygons, the surface associated with each polygon and the chart of mean annual precipitation record in the period 1965–2005.

1.1.2 Inverse Distance Weighted Interpolation (IDW)

Inverse Distance Weighted Interpolation (IDW) is a method based on the assumption that the value of the target variable in a new location is strongly

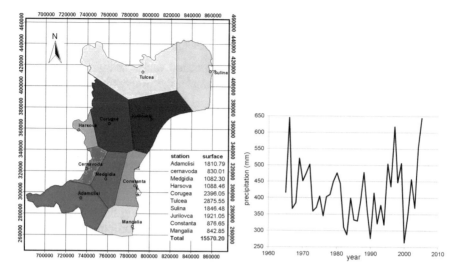

Fig. 1 Thiessen polygons for the annual precipitation series (1965–2005) record in Dobrogea

influenced by the values from the nearest points and is less influenced by those from the locations situated at a higher distance. Therefore, the weight with which the values from the neighbor locations participate in the determination of the value in the study location is greater if the locations are closer to the study point.

Essentially, all the spatial interpolation methods compute weighted averages of sampled data as estimations for unknown data [17]. Therefore, the solution of the interpolation problem is:

$$\hat{z}(s_0) = \sum_{i=1}^{n} \lambda_i(s_0) \cdot z(s_i), \tag{1}$$

where $\hat{z}(s_0)$ is the value estimated for the variable of interest, Z, at the station s_0, $z(s_i)$ is the sample value at the station s_i, $\lambda_i(s_0)$ is the weight corresponding to the station s_i, n is the number of stations and $\sum_{i=1}^{n} \lambda_i(s_0) = 1$.

The main difference among the spatial interpolation methods lies in computing the weights $\lambda_i(s_0)$ used in the interpolation [20].

The simplest modality of weights' determination in IDW is given by the formula [25]:

$$\lambda_i(s_0) = \frac{1/d^\beta(s_0, s_i)}{\sum_{i=1}^{n} (1/d^\beta(s_0, s_i))}, \quad \beta > 1, \tag{2}$$

where $d(s_0, s_i)$ is the distance from s_0 to s_i and β is a coefficient that adjusts the weights.

Choosing the parameter β is an optimization process by itself. Usually, the search for the optimal β is a grid search: a specific range is set (arbitrarily, or based on the researcher's intuition) and then β traverses all the values in that range, with a certain step-size. The value yielding the lowest prediction error among the searched values is attributed to the parameter.

Other versions of IDW computes the interpolated values using only the closest neighbors in a R-disc, instead of the entire sample, the weights being in this case:

$$\lambda_i(s_0) = \left(\frac{R - d(s_0, s_1)}{Rd(s_0, s_1)} \right)^2 \tag{3}$$

A modified IDW is investigated in [24], where the elevation is also taken into account for the estimation of values at unknown locations. An attempt to find the search radius and the power parameter of IDW employing a genetic algorithm is reported in [8]. A genetic algorithm is used in [7] for finding the optimal β.

An adaptive version is proposed in [19], where the algorithm that selects the weights considers the stations' density in the neighborhood of the study location.

Remark that IDW cannot produce higher values than the maxima registered in the sampling points because the stations' influence decreases with the distance to a given point.

1.1.3   Nearest Neighbor Interpolation

In the nearest neighbor interpolation method, the value at the study point is equal to the values at the closest neighbor station, the values at the other stations being neglected. Therefore, the results is an interpolant piecewise constant.

1.1.4   Natural Neighbor Interpolation

Thiessen polygons method is the base of the natural neighbor interpolation method [26]. Here, the weights are computed using the local coordinates, i.e. the coordinates of the points that contribute to the estimation of the value at the study point. The algorithm determines two partitions of the region in polygons. The first one contains the interpolation point and the second one, not. The interpolation is carried out by including in computation only those points whose zones of influence were altered by the elimination of the interpolation point [16].

1.1.5   Splines și Local Trend Surface

A spline is a curve built by polynomial segments subject to some continuity conditions in knots. Spline functions are determinist interpolators that can be used to adjust a surface through each set of data, minimizing at the same time its curvature [11, 12].

Unlike IDW, the splines can estimate values greater than the registered maxima or less than the registered minima in the sample points. Splines functions generate the surfaces by calculating weighted averages of the neighboring values and passing through the known locations.

Different types of splines are used, as, for example:

- The cubic spline [22],
- The regularized spline:

$$\varphi(r) = -\sum_{n=1}^{\infty} \frac{(-1)^n (\lambda r)^{2n}}{n!n} = \ln(\lambda r/2)^2 + E_1(x) + C_E,$$

where ln is the natural logarithm, C_E is the Euler's constant and $E_1(x) = \int_{-\infty}^{x} e^t/t dt$;

- Spline with tension:

$$\varphi(r) = \ln(\lambda r/2)^2 + K_0(\lambda r) + C_E,$$

where $K_0(x)$ is the modified Bessel function;

- Multi-quadratic spline, $\varphi(r) = \sqrt{\lambda^2 + r^2}$;
- Multi-quadratic inverse spline: $\varphi(r) = 1/\sqrt{\lambda^2 + r^2}$;
- *Thin Plate Spline*: $\varphi(r) = (\lambda r)^2 \ln(\lambda r)$.

The smoothing parameter, λ, is utilized for determining the surface unevenness and is estimated by minimizing the mean square prediction errors, using the cross-validation procedure [14]. The higher the smoothing parameter is, the smoother the surface is, excepting for the multi-quadratic splines.

In the *Thin Plate Splines* (TPS) (*Laplacian smoothing splines*) method, developed for climatic data [28], the smoothing parameter is determined by the minimization of the generalized cross-validation function [17].

For an overview of the mathematical formalism of the splines smoothing and a comparison with the kriging, the readers may refer to [21].

Local trend surfaces (LTS) are built by a polynomial approximation for each target point, using neighbor samples. Two approaches to this technique are known: the local polynomial regression and the bilinear or bicubic splines. For the first one, the readers may refer to [10] and for the second one, to [1, 2].

1.2 Statistical Methods

1.2.1 Variogram

The spatial correlation is modeled in geostatistics by variogram, that shows the dependence of the semivariances on distances. If in the classical statistics problems, the correlation is estimated by the correlation coefficient or by using a scatterplot, the spatial correlation between the observations of the same variable recorded in different locations can be estimated in conditions related to the number of pairs of the recorded values and to the stationarity of the process that generates them [13, 15].

Cressie [11] defines the variogram as the variance of the difference between the values of a random field Z at two locations, x and y, across the field realizations, i.e.:

$$2\gamma(h) = Var[Z(x) - Z(y)].$$

In the hypothesis that the random field is stationary, the variogram of Z is defined by:

$$2\gamma(h) = E\{[Z(x) - Z(y)]^2\} = E\{[Z(x) - Z(x+h)]^2\},$$

where $E(Z(x))$ is the expectance of $Z(x)$ and h is the distance from x to y.

Therefore, in the stationarity hypothesis, one can form pairs $\{z(s_i), z(s_j)\}$ that have the same separation vector $h = s_i - s_j$ and estimate the correlations.

If the process is isotropic, i.e. $\gamma(h)$ does not depend on the direction, the vector h can by replaced by its length, in the sense of Euclidean norm. Therefore, the variogram can be is estimated by the sample variogram, defined by:

$$2\tilde{\gamma}(h_j) = \frac{1}{N_h} \sum_{i=1}^{N_h} (z(s_i) - z(s_i + h))^2, \quad \forall h \in \tilde{h}_j,$$

where s_i is the location i, N_h is the number of the samples of data pairs $(z(s_i), z(s_i + h))$ and $\tilde{h}_j$ is the family of distances [15].

A large class of models is obtained when the mean of the process has a spatial variation, and the process can be represented as a linear function of known predictors, $X_j(s)$, as in the equation:

$$Z(s) = \sum_{j=1}^{p} X_j(s)\beta_j + e(s) = X\beta + e(s), \tag{4}$$

where $X_j(s)$ are known spatial regressors, β_j are the unknown regression coefficients, $e(s)$ is the residual, $X_0(s) \equiv 1$, X is a $n \times (p+1)$ matrix whose columns are $X_j(s)$ and β is a column vector, containing the $p+1$ unknown coefficients.

For this type of models, the stationarity properties concern the residual, such that the sample variogram must be computed for the estimated residual [6].

There are two simple modalities of analysis of the spatial correlations. The first one is the use of the scatterplot of the pairs $(z(s_i), z(s_j))$ grouped function of the distance between them, $h_{ij} = d(s_i, s_j)$. The second one is plotting the variogram and the points cloud obtained by representing the pairs $(z(s_i) - z(s_j))^2$ function of their separation distances, h_{ij}.

The following parameters characterize a variogram:

- *Sill (s)*—the semivariance value at which the variogram levels off or the limit of the variogram when lag tends to infinity.
- *Range (r)*—the lag at which the semivariogram reaches the sill value.
- *Nugget (a)*—In theory, at lag 0 (i.e. at the origin), the value of the semivariogram value should be zero. If for lags close to zero, it significantly differs from zero, then this value is called Nugget [23].

For reasons related to the kriging applications (as the requirement that the semivariogram must be non-negative defined), the empirical semivariogram is replaced in computation by different semivariogram models. The most frequent used are [3, 15]:

- Gaussian:

$$\gamma(h) = \begin{cases} 0, & h = 0 \\ a + (s - a)\left[1 - \exp\left(\frac{-3|h|^2}{a^2}\right)\right], & h \neq 0 \end{cases}, \quad s \geq a \geq 0, \ r \geq 0$$

- Spherical:

$$\gamma(h) = \begin{cases} 0, & h = 0 \\ a + (s - a)\left[\frac{3h}{2a} - \frac{h^3}{2a^3}\right], & 0 < |h| < a, \quad s \geq a \geq 0, \ r \geq 0, \\ s, & |h| \geq a \end{cases}$$

- Exponential:

$$\gamma(h) = \begin{cases} 0, & h = 0 \\ a + (s - a)\left[1 - \exp\left(\frac{-3|h|}{a}\right)\right], & h \neq 0 \end{cases}, \quad s \geq a \geq 0, \ r \geq 0.$$

In this model, the sill is reached only asymptotically (for $|h| \to \infty$) and the range is equal to the distance at which the relation $\gamma(h) = 0.95s$ is fulfilled.

The distance at which the tangent at zero crosses the sill is equal to $1/3r$.

- Power:

$$\gamma(h) = \begin{cases} 0, & h = 0 \\ a + b|h|^p, & h \neq 0 \end{cases}, \quad 0 \leq p < 2, \ a, b \geq 0.$$

A particular case is the linear one, obtained for $p = 1$.

1.2.2   Kriging

Kriging (KG) is the generic name of a family of generalized least-squares regression algorithms which uses the variogram [17]. Employing this method, one wants to find the best linear estimation of the average value at a location, utilizing the values recorded in the neighbor locations. The weights attached to the known values are selected such that to minimize the variance of the resulted estimation, taking into account the geometry of the observation points and the spatial variability [6].

Different types of kriging have been developed, as ordinary, universal, with trend, factorial, dual, block kriging, etc. that may be used function of the problem at hand. For an extensive study of this topic we recommend [6, 9, 11].

Examples of packages dedicated to spatial data interpolation in R are: **akima, deldir, fields, geoR, GeoRglm, GRASS, gstat, spatial, sgeostat, RandomFields, tripac**.

In the following we shortly refer to some kriging versions, based on [6].

As mentioned, the problem of spatial prediction refers to the estimation of unknown quantities $Z(s_0)$ based on some hypotheses about the type of the trend of the random field Z, its variance and the spatial correlation, using some data samples, $z(s_i)$.

Suppose that the trend can be described by a linear regression function as in (4), $x(s_0)$ is a $(1 \times p)$ matrix that contains the values of the predictor for s_0, V is the covariance matrix of $Z(s)$ and v is the covariance vector of $Z(s)$ and $Z(s_0)$. Then, the best unbiased predictor of $Z(s_0)$ is:

$$\hat{Z}(s_0) = x(s_0)\hat{\beta} + v'V^{-1}(Z(s) - X\hat{\beta}) = (x(s_0) - v'V^{-1}X)\hat{\beta} + v'V^{-1}Z(s), \quad (5)$$

where

$$\hat{\beta} = (X'V^{-1}X)^{-1}X'V^{-1}Z(s)$$

is the least squares generalized estimator of the trend coefficients and X' is the transposed of the X matrix.

The weights $v'V^{-1}$ in (5) are called *weights of the simple kriging*.

The variance of the mean error of the predictor $\hat{Z}(s_0)$ is given by:

$$\sigma^2(s_0) = \sigma_0^2 - v'V^{-1}v + \delta(X'V^{-1}X)^{-1}\delta', \quad (6)$$

where σ_0^2 is the variance of $Z(s_0)$ and $\delta = x(s_0) - v'V^{-1}X$.

If all observations are not correlated with $Z(s_0)$, then $v = 0$ and $v'V^{-1}v = 0$.

It has been proved that $(X'V^{-1}X)^{-1} = Var(\hat{\beta} - \beta)$, so it vanishes if s_0 is one of the observed locations.

If the number of predictors is greater than or egal to 1, we speak about the *universal kriging*.

If $p = 1$ and X doesn't include coordinates, we speak about *kriging with external drift*.

If the mean of the process $Z(s)$ is constant, then $p = 0$ and $X_0 = 1$, and we discuss about *ordinary kriging*.

Simple kriging is obtained when β is known. In this case $\hat{\beta}$ is replaced in (5) by β and the variance is obtained by omitting the third term in (6).

2 Applications of the Spatial Interpolation Methods

2.1 The Most Probable Precipitation Method

Given the series of precipitations recorded at k locations, at the same time $t(t = \overline{1,n})$, denote by $(y_{it})_{t=\overline{1,n}}$ the series registered at the station $i(i = \overline{1,k})$.

In [4] we introduced the most probable precipitation (MPPM) for interpolation the regional precipitation, that will be applied in the following. It has the steps:

1. Build a $(n \times k)$ matrix such that the column i contains the precipitation recorded at the station i.
2. Compute the difference between the minimum $y_{j\,\min}$ and the maximum $y_{j\,\max}$ of the elements from each row of the matrix, and denote the result by $A_j, j = \overline{1,n}$.
3. Divide the intervals $[y_{j\,\min}, y_{j\,\max}]$ in sub-intervals of the same length L_j and denote their number by $n_j \in N^*$. The number of values in an interval is called the interval's frequency.
4. Build couples formed by the sub-interval and the corresponding frequency, i.e. (I_{kj}, f_{kj}), $k = \overline{1,n_j}$, $j = \overline{1,n}$, where I_{kj} is the k sub-interval of the period j, and f_{kj} is the frequency of I_{kj} .
5. Choose the sub-interval/sub-intervals I_{kj} with the maximum frequency, $j = \overline{1,n}$. If there is only one interval with this property, denote it and its frequency respectively by $I_{j\,\max}$ and $f_{j\max}$ and go to 7. If there are many such sub-intervals, go to 6.
6. Calculate the average precipitation registered at all the stations and the average precipitation registered in the same period at the stations whose values are in the sub-intervals with the same frequency. Define $I_{j\,\max}$ to be the sub-interval whose average precipitation is closer to the average precipitation registered in the entire period j. Pass to 7.
7. Choose the regional precipitation estimation for the period j to be the average of the values from $I_{j\,\max}$.

We apply MPPM to the precipitation series registered in Spring in the period 1965–2005 at the main meteorological stations from Dobrogea, but Sulina. The results are presented in Tables 1 and 2 only for the first 39 months. Table 1 contains: the months, the sub-intervals' length for $n_j = n = 3$, $j = \overline{1,9}$ and the sub-intervals. Table 2 contains the frequency of each interval, the mean monthly precipitation and the regional precipitation estimation.

Table 1 The amplitudes, A_j, the sub-intervals and their lengths (mm) in MPPM

Month	A_j	L_j	I_{1j}	I_{2j}	I_{3j}
Mar-65	21.4	7.1	[11.3, 18.4)	[18.4, 25.6)	[25.6, 32.7]
Apr-65	46.0	15.3	[0.6, 15.9)	[15.9, 31.3)	[31.3, 46.6]
May-65	85.9	28.6	[15.7, 44.3)	[44.3, 73.0)	[73.0, 101.6]
Mar-66	31.4	10.5	[15.3, 25.8)	[25.8, 36.2)	[36.2, 46.7]
Apr-66	33.6	11.2	[11.4, 22.6)	[22.6, 33.8)	[33.8, 45.0]
May-66	30.9	10.3	[16.5, 26.8)	[26.8, 37.1)	[37.1, 47.4]
Mar-67	18.3	6.1	[12.2, 18.3)	[18.3, 24.4)	[24.4, 30.5]
Apr-67	12.7	4.2	[9.9, 14.1)	[14.1, 18.4)	[18.4, 22.6]
May-67	32.6	10.9	[13.7, 24.6)	[24.6, 35.4)	[35.4, 46.3]
Mar-68	95.9	32.0	[12.1, 44.1)	[44.1, 76.0)	[76.0, 108]
Apr-68	5.7	1.9	[0.0, 1.9)	[1.9, 3.8)	[3.8, 5.7]
May-68	8.9	3.0	[0.0, 3.0)	[3.0, 5.9)	[5.9, 8.9]
Mar-69	17.3	5.8	[23.0, 28.8)	[28.8, 34.5)	[34.5, 40.3]
Apr-69	29.2	9.7	[22.8, 32.5)	[32.5, 42.3)	[42.3, 52.0]
May-69	30.2	10.1	[3.5, 13.6)	[13.6, 23.6)	[23.6, 33.7]
Mar-70	25.5	8.5	[22.6, 31.1)	[31.1, 39.6)	[39.6, 48.1]
Apr-70	14.1	4.7	[10.6, 25.3)	[15.3, 20.0)	[20.0, 24.7]
May-70	94.5	31.5	[72.4, 103.9)	[103.9, 135.4)	[135.4, 166.9]
Mar-71	13.2	4.4	[23.1, 27.5)	[27.5, 31.9)	[31.9, 36.3]
Apr-71	13.1	4.4	[3.8, 8.2)	[8.2, 12.5)	[12.5, 16.9]
May-71	190.9	63.6	[17.2, 80.8)	[80.8, 144.5)	[144.5, 208.1]
Mar-72	6.5	2.2	[2.8, 5.0)	[5.0, 7.1)	[7.1, 9.3]
Apr-72	19.4	6.5	[11.5, 18)	[18.0, 24.4)	[24.4, 30.9]
May-72	44.7	14.9	[20.9, 35.8)	[35.8, 50.7)	[50.7, 65.6]
Mar-73	31.3	10.4	[51.7, 67.5)	[67.5, 78.0)	[78.0, 84.4]
Apr-73	57.8	19.3	[21.0, 40.3)	[40.3, 59.5)	[59.5, 78.8]
May-73	76.8	25.6	[24.6, 50.2)	[50.2, 75.8)	[75.8, 101.4]
Mar-74	14.1	4.7	[10.6, 15.3)	[15.3, 20.0)	[20.0, 24.7]
Apr-74	44.5	14.8	[35.0, 49.8)	[49.8, 64.7)	[64.7, 79.5]
May-74	39.7	13.2	[25.3, 38.5)	[38.5, 51.8)	[51.8, 65.0]
Mar-75	25.8	8.6	[14.0, 22.6)	[22.6, 31.2)	[31.2, 39.8]
Apr-75	53.9	18.0	[23.7, 41.7)	[41.7, 59.6)	[59.6, 77.6]
May-75	45.4	15.1	[17.3, 32.4)	[32.4, 47.6)	[47.6, 62.7]
Mar-76	6.9	2.3	[8.1, 10.4)	[10.4, 12.7)	[12.7, 15.0]
Apr-76	21.2	7.1	[9.1, 16.2)	[16.2, 23.2)	[23.2, 30.3]
May-76	21.1	7.0	[8.4, 15.4)	[15.4, 22.5)	[22.5, 29.5]
Mar-77	21.4	7.1	[1.0, 9.3)	[9.3, 17.7)	[17.7, 26.0]
Apr-77	46.0	15.3	[24.9, 46.4)	[46.4, 67.8)	[67.8, 89.3]
May-77	85.9	28.6	[26.1, 56.1)	[56.1, 86.2)	[86.2116.2]

Table 2 The sub-intervals' frequencies, the mean monthly precipitation and the precipitation estimated by MPPM

Month	f_{1j}	f_{2j}	f_{3j}	Mean	Estimation
Mar-65	3	2	4	21.58	29.5
Apr-65	1	5	3	29.41	27.0
May-65	1	2	6	75.64	91.6
Mar-66	2	3	4	33.33	42.2
Apr-66	2	3	4	29.73	39.6
May-66	4	1	4	30.27	30.3
Mar-67	3	5	1	19.60	21.2
Apr-67	3	5	1	15.20	16.4
May-67	4	1	4	31.20	31.2
Mar-68	8	0	1	26.79	16.6
Apr-68	3	4	2	2.42	2.7
May-68	6	1	2	2.83	0.6
Mar-69	4	1	4	31.63	31.6
Apr-69	7	1	1	30.56	26.1
May-69	6	1	2	13.94	7.6
Mar-70	6	2	1	28.77	24.1
Apr-70	1	4	4	18.86	16.9
May-70	2	4	3	124.61	117.4
Mar-71	4	3	2	28.70	24.7
Apr-71	3	4	2	9.62	10.6
May-71	4	2	3	111.06	58.2
Mar-72	6	2	1	4.71	3.7
Apr-72	3	5	1	20.00	21.6
May-72	3	5	1	39.72	41.3
Mar-73	3	2	4	73.36	83.9
Apr-73	4	2	3	50.03	30.6
May-73	7	1	1	42.12	31.9
Mar-74	4	0	5	17.46	22.6
Apr-74	4	2	3	55.17	44.8
May-74	6	2	1	38.81	31.5
Mar-75	4	3	2	25.94	17.6
Apr-75	4	2	3	48.66	29.7
May-75	3	3	3	39.08	40.1
Mar-76	2	3	4	12.37	14.4
Apr-76	4	3	2	18.92	12.2
May-76	3	3	3	18.70	18.9
Mar-77	5	2	2	10.83	4.4
Apr-77	3	3	3	58.31	45.5
May-77	4	4	1	61.39	66.3

Fig. 2 Estimation of regional
precipitation in Spring (1965–
2005) obtained by the
Thiessen polygons method
and MPPM

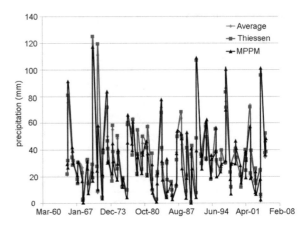

The performances of this method are evaluated using the mean absolute error
(MAE) and the mean absolute percentage error (MAPE), defined respectively by:

$$\text{MAE} = \frac{1}{n}\sum_{t=1}^{n}|e_t| \quad \text{and} \quad \text{MAPE} = \frac{1}{n}\sum_{t=1}^{n}\frac{|e_t|}{y_t} \times 100,$$

where $|e_t|$ is the absolute prediction error at the moment t and y_t is the value
recorded at the same moment.

For the same data series, the Thiessen polygons method, IDW and the ordinary
kriging (OK) have been used for the precipitation estimation.

The chart from Fig. 2 is built using the values estimated by the Thiessen
polygons method and MPPM for the regional precipitation in Spring, together with
the average precipitation registered in the same season in 1965–2005.

We remark that the values obtained by MPPM are closer to those of the mean
regional precipitation than those obtained by the Thiessen polygons method.

The comparison of the performances of these two methods with respect to MAE
and MAPE are presented in Table 3.

The overall MAE corresponding to the Thiessen polygons method is 9.5 and to
MPPM is 10.0. The overall MAPE corresponding to the Thiessen polygons method
is of 52.7 and to MPPM is 44.0. So, MPPM performs better than the Thiessen
polygons method in this case.

Table 4 contains the values of MAE and MAPE from the kriging (with the
exponential variogram model) and IDW interpolation with $\beta = 2$.

The overall MAEs are respectively 10.5 in kriging and 10.7 in IDW. The overall
MAPEs are respectively 56.1, in kriging and 50.2 in IDW.

Concluding, with respect to MAE, the best results have been obtained using the
Thiessen polygons method, but with respect to MAPE, the best ones are those given
by MPPM.

Table 3 MAE and MAPE in Thiessen polygons methods and MPPM for the precipitation estimation in Spring (1965–2005)

Series	Adamclisi	Cernavoda	Constanta	Corugea	Harsova	Jurilovca	Mangalia	Medgidia	Tulcea
MAE									
MPPM	11.80	10.70	9.80	10.5	9.60	9.20	10.60	7.60	11.2
Thiessen	10.01	9.50	10.05	8.29	10.34	10.09	10.36	7.68	9.17
MAPE									
MPPM	30.7	30.8	72.0	45.2	44.1	57.5	53.2	22.3	40.3
Thiessen	33.8	35.7	90.6	44.7	52.5	70.5	78.0	30.8	37.5

Table 4 MAE and MAPE in IDW and ordinary kriging for the precipitation in Spring

Series	Adamclisi	Cernavoda	Constanta	Corugea	Harsova	Jurilovca	Mangalia	Medgidia	Tulcea
MAE									
Kriging	10.0	10.5	9.9	9.9	10.5	8.4	13.1	10.8	11.0
IDW	9.8	10.0	10.0	9.7	9.7	8.1	17.8	9.9	11.8
MAPE									
Kriging	30.9	42.6	89.9	50.8	78.2	32.7	61.0	75.4	43.1
IDW	30.0	30.0	88.6	46.9	70.9	31.3	50.2	63.5	40.2

2.2   Spatial Interpolation of Maximum Annual Precipitation

The results of the spatial interpolation of the maximum annual precipitation series by ordinary kriging and IDW are presented in this section. We worked with the data series recorded in the period 1965–2005 at the ten main meteorological series from Dobrogea.

2.2.1   Spatial Interpolation by Kriging

The exponential variogram model has been used in the spatial interpolation done by kriging. The charts from Figs. 3 and 4 are drawn based on the interpolated values of Corugea and Adamclisi maximum annual precipitation series.

The curves are denoted respectively by: *predicted* 9, when the other nine series have been utilized, and *predicted* 8, when using the same series, but Sulina.

As goodness of fit indicators, MSE and MAE are reported (Table 5).

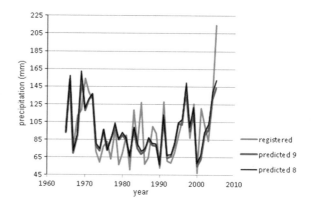

Fig. 3 Interpolation of maximum annual precipitation series at Corugea (1965–2005)

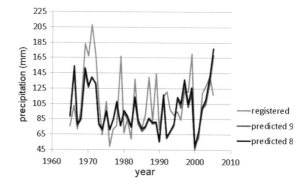

Fig. 4 Interpolation of maximum annual precipitation series at Adamclisi (1965–2005)

Table 5 MSEs and MAEs in ordinary kriging interpolation

Series	MSE		MAE	
	Predicted 9	Predicted 8	Predicted 9	Predicted 8
Adamclisi	32.73	32.60	25.09	24.88
Cernavoda	22.11	23.32	16.08	17.43
Constanta	30.25	30.03	18.63	19.39
Corugea	22.51	22.04	16.20	16.67
Mangalia	41.97	41.63	26.48	27.15
Medgidia	23.86	22.86	20.11	19.33
Harsova	35.91	36.82	26.13	27.45
Jurilovca	22.67	26.73	18.63	21.56
Tulcea	33.85	32.65	26.56	25.36

MSEs are comparable, being smaller in 6 cases for *predicted* 8. MAEs are smaller only in 3 cases for *predicted* 8. The estimated values of the series recorded at the stations situated near the region boundary (Mangalia, Sulina) are worse estimated.

The values estimated for Sulina series are represented in Fig. 5. The chart of the interpolated values has the same shape as that of the recorded data, but the estimation error is significant, as expected, because Sulina doesn't belong to the homogenous group formed by the other nine series. The mean standard error obtained in this estimation is 43.84 and MAE is 35.95.

Table 6 contains MSEs and MAEs from the wavelet model for the estimation of the maximum regional precipitation [5].

MSE and MAE in the kriging interpolation are smaller than those in the wavelets model in the majority of cases, so the kriging performs better on the study data.

Fig. 5 Interpolation of maximum annual precipitation series at Sulina (1965–2005)

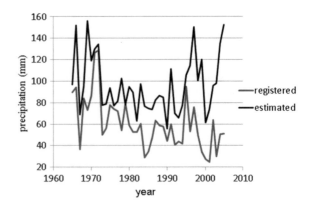

Table 6 MSEs and MAEs in the wavelets model for annual maximum precipitation

	Adamclisi	Cernavoda	Constanta	Corugea	Harsova	Jurilovca	Mangalia	Medgidia	Sulina	Tulcea
MSE	32.76	31.43	34.71	38.46	35.40	35.97	43.82	26.98	36.10	25.10
MAE	24.63	21.76	22.63	19.50	26.34	19.83	22.46	21.57	30.59	19.15

2.2.2 Spatial Interpolation by IDW

For comparison with the results from the previous section, we present the minimum mean standard errors and MAE in IDW interpolation of the maximum annual precipitation, when the parameter β passes through the interval (1, 4] with a step of 0.1. The values of β for which the minima are reached are also indicated in Table 7.

MAEs and MSEs in IDW interpolation are smaller than those from the wavelets model in 70 % of cases, so for the study case, IDW performs better than the wavelets method.

To determine the influence of an additional quantity of information on the interpolation results, we took into consideration 41 secondary series together with the main ones. The values of $\beta \in$ (1, 4] that minimizes MAEs and MSEs in this case are given in Table 8. Note that β has been selected by a grid search with the step of 0.1.

Analyzing Table 8 it results that:

- Introducing new series in interpolation didn't lead to MSEs' decrease. If for Adamclisi, Cernavoda, Constanta and Corugea, the values of MSEs were comparable with those got using only the main series, for Harsova, Medgidia and Sulina the MSEs increased, but for Jurilovca, Mangalia și Tulcea they decreased. MSEs varied in tight limits, the highest being registered at Sulina (45.26).
- The results of IDW are in concordance with the hypothesis that there exists a moderate correlation between the values of precipitation series registered in Dobrogea, in general, and those of the extremes ones, in particular.

Finally a common value of the parameter β that minimizes the total MSE in IDW interpolation for the maximum annual precipitation at the main series has been determined to be equal to 1.1934.

The search of β has also been done by a grid search in the interval (1, 4] with a step of 0.0001. In Table 9 we present the MSEs and MAEs corresponding to the main stations, in IDW interpolation for $\beta = 1.1934$. Remark that they are comparable with those obtained by the kriging interpolation.

2.3 Spatial Interpolation of the Parameters in the AR Models for the Annual Precipitation Series

In this section we discuss the spatial interpolation of the parameters of AR models for the annual precipitation series registered at the main meteorological series and we compare the results obtained by different methods. We also address the issue of the influence of supplementary information on the parameters' estimation, also taking into account the series registered at the 41 secondary stations in Dobrogea. Finally, we want to answer to the following question. Are these methods appropriated for estimation the parameters of AR(p) models for ungauged locations, models that, at their turn are used for the estimation of precipitation at those locations?

Table 7 The values of the parameter β that minimizes MSE and MAE in IDW for the maximum annual precipitation series registered at the main stations

	Adamclisi	Cernavoada	Constanta	Corugea	Harsova	Jurilovca	Mangalia	Medgidia	Sulina	Tulcea
β	4	1.1	1.1	1.1	1.1	1.1	1.1	1.1	2.6	1.1
MSE	28.64	22.55	29.96	22.21	35.11	23.88	41.65	22.43	43.25	36.61
MAE	22.20	16.82	19.34	16.50	25.30	19.53	26.73	19.1 7	36.20	29.16

Table 8 The values of the parameter β that minimizes MSE and MAE in IDW for the maximum annual precipitation series registered at all 51 stations from Dobrogea

	Adamclisi	Cernavoda	Constanta	Corugea	Harsova	Jurilovca	Mangalia	Medgidia	Sulina	Tulcea
β	2	1.2	2.4	1.2	2.6	1.8	2.6	1.2	2.6	3.4
MSE	28.2	22.5	29.9	22.4	40.0	21.8	32.1	25.3	45.3	23.1
MAE	18.9	16.3	19.3	19.9	25.8	18.2	24.3	20.1	37.7	16.8

	Agigea	Albești	Altân Tepe	Amzacea	Baia	Băltăgești	Biruința	Casian	Casimcea	Ceamurlia
β	2.4	4	1.6	2	1.2	2.8	2.6	1.6	2.6	1.6
MSE	32.16	39.65	26.88	23.93	26.00	30.32	26.79	27.06	28.72	24.8
MAE	19.02	27.56	21.26	17.76	19.22	21.39	28.13	21.90	24.69	18.1

	Cerna	Cheia	Cobadin	Corbu	Crucea	Cuza Vodă	Dăeni	Dobromir	Dorobanțu	Dorobanțu II
β	3.4	1.6	1.4	1.4	2	2.6	4	1.6	1.1	2.6
MSE	23.88	19.59	28.71	25.03	29.82	23.46	40.18	28.09	33.84	34.31
MAE	18.87	14.49	20.60	18.49	22.23	16.57	27.04	21.87	22.60	23.25

	Greci	Hamcearca	Independența	Lipnița	Lumina	Mihai Viteazu	Negru Vodă	Niculițel	Nuntași	Pantelimon
β	3.4	4	2.8	1.4	2.8	1.1	2.0	1.1	1.4	1.8
MSE	23.02	25.47	24.71	27.81	24.88	28.77	35.61	44.07	27.22	33.01
MAE	16.79	17.71	19.15	19.57	18.04	21.30	22.33	34.54	19.22	24.86

	Peceneaga	Pecineaga	Peștera	Pietreni	Posta	Săcele	Saraiu	Satu Nou	Siliștea	Topolog	Zebril
β	1.5	2	1.6	2.4	2	1.4	2.4	1.1	1.6	2	2
MSE	24.15	36.0	36.20	31.71	23.65	21.64	23.81	23.40	22.78	30.70	27.75
MAE	17.45	22.2	26.98	23.84	16.54	17.56	18.72	18.01	16.03	22.41	21.75

Table 9 MSEs and MAEs in IDW with β = 1.1934

	Adamclisi	Cernavoda	Constanta	Corugea	Harsova	Jurilovca	Mangalia	Medgidia	Sulina	Tulcea
MSE	32.42	22.91	30.32	22.41	35.63	24.10	42.04	22.38	44.02	33.08
MAE	24.83	17.16	19.66	16.65	26.17	19.57	26.78	19.05	36.31	25.93

Fig. 6 Interpolation of AR
parameters by the nearest
neighbor method

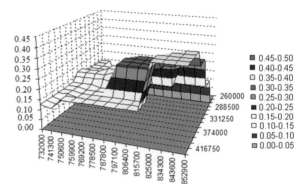

Fig. 7 Interpolation of AR
parameters by the natural
neighbor method

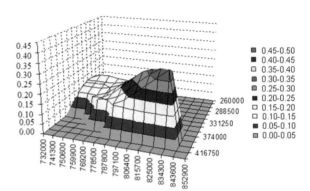

The results of the spatial interpolation of the parameters of AR models for the series recorded at the main ten meteorological stations by the nearest neighbor and the natural neighbor methods are represented in Figs. 6 and 7.

Analyzing Fig. 6 and Table 10 it results that more than a third of the parameters of the possible AR models are −0.141, 0.307 and 0.388, that are the extreme values of the parameters in AR(1) models for the main series, fact that is not in concordance with the reality. So, the nearest neighbor method doesn't give satisfactory results.

Even if the natural neighbor method doesn't provide parameters' estimations for all the locations whose coordinates are given in the grid (Table 11) and there is a certain spatial homogeneity of the distribution of the estimated parameters, the values provided by it are closer to those from the AR(1) models. Since the spatial distribution of the stations is not homogenous and the stations' density is not high, the expected results could not be very good.

The results of spatial interpolation of the AR(1) parameters in the models for the 51 series of annual precipitation (10 main + 41 secondary) are represented in Figs. 8 and 9 and the interpolation values are given in Tables 12 and 13.

Table 10 Spatial interpolation of the parameters in the AR models for the main series by the nearest neighbor method

Coordinate	263,000	275,800	288,600	301,400	314,200	327,000	339,800	352,600	365,400	378,200	391,000	403,800	416,600
733,000	0.142	0.142	0.142	0.142	0.169	0.169	0.127	0.127	0.127	0.127	0.127	0.127	0.127
741,700	0.142	0.142	0.142	0.142	0.169	0.169	0.169	0.127	0.127	0.127	0.127	0.115	0.115
750,400	0.142	0.142	0.142	0.142	0.169	0.169	0.169	0.127	0.115	0.115	0.115	0.115	0.163
759,100	−0.141	0.142	0.142	0.187	0.187	0.187	0.169	0.115	0.115	0.115	0.115	0.115	0.163
767,800	−0.141	−0.141	0.187	0.187	0.187	0.187	0.115	0.115	0.115	0.115	0.115	0.163	0.163
776,500	−0.141	−0.141	0.146	0.146	0.146	0.187	0.115	0.115	0.115	0.115	0.115	0.163	0.163
785,200	−0.141	−0.141	0.146	0.146	0.146	0.146	0.146	0.115	0.115	0.307	0.163	0.163	0.163
793,900	−0.141	−0.141	0.146	0.146	0.146	0.146	0.146	0.307	0.307	0.307	0.307	0.163	0.163
802,600	−0.141	−0.141	0.146	0.146	0.146	0.146	0.307	0.307	0.307	0.307	0.307	0.163	0.163
811,300	−0.141	−0.141	0.146	0.146	0.146	0.146	0.307	0.307	0.307	0.307	0.307	0.163	0.163
820,000	−0.141	−0.141	0.146	0.146	0.146	0.146	0.307	0.307	0.307	0.307	0.307	0.163	0.163
828,700	−0.141	−0.141	0.146	0.146	0.146	0.146	0.307	0.307	0.307	0.307	0.388	0.388	0.388
837,400	−0.141	−0.141	0.146	0.146	0.146	0.146	0.307	0.307	0.307	0.307	0.388	0.388	0.388
846,100	−0.141	−0.141	0.146	0.146	0.146	0.307	0.307	0.307	0.307	0.388	0.388	0.388	0.388

Table 11 Spatial interpolation of the parameters in the AR models for the main series by the natural neighbor method

Coordinate	263,000	275,800	288,600	301,400	314,200	327,000	339,800	352,600	365,400	378,200	39,100	403,800	416,600
733,000	–	–	–	–	–	–	–	–	–	–	–	–	–
741,700	–	–	–	0.148	0.156	0.159	0.148	0.132	–	–	–	–	–
750,400	–	–	–	0.147	0.174	0.163	0.147	0.131	0.126	–	–	–	–
759,100	–	–	0.059	0.136	0.181	0.160	0.145	0.130	0.120	0.138	–	–	–
767,800	–	–	0.046	0.124	0.172	0.159	0.152	0.143	0.129	0.138	–	–	–
776,500	–	−0.047	0.034	0.114	0.162	0.163	0.166	0.166	0.164	0.150	0.149	–	–
785,200	–	−0.059	0.023	0.105	0.157	0.173	0.186	0.195	0.201	0.183	0.162	–	–
793,900	–	–	−0.017	0.069	0.145	0.188	0.210	0.226	0.240	0.225	0.194	0.168	–
802,600	–	–	–	0.001	0.081	0.153	0.213	0.253	0.279	0.266	0.235	0.205	–
811,300	–	–	–	–	–	0.102	0.174	0.239	0.291	0.306	0.275	0.245	–
820,000	–	–	–	–	–	–	–	0.201	0.267	0.317	0.314	0.284	–
828,700	–	–	–	–	–	–	–	–	–	0.297	0.343	0.323	–
837,400	–	–	–	–	–	–	–	–	–	–	0.329	0.362	–
846,100	–	–	–	–	–	–	–	–	–	–	–	–	–

Fig. 8 Interpolation of AR parameters by the nearest neighbor method using 51 annual series

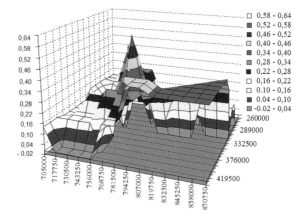

Fig. 9 Interpolation of AR parameters by natural neighbor methods using 51 annual series

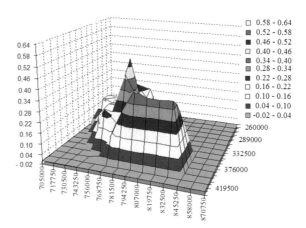

The results of the estimations registered an improvement, even if there still are over-estimations of the coefficients corresponding to the AR models for the sites situated in the neighborhood of the stations whose coefficients have high values, and, as consequence, values the precipitation estimated in those locations based on the new AR models (with the estimated coefficients) are not satisfactory.

Table 12 Results of parameters' interpolation in AR models using 51 series by the natural neighbor method precipitation series

Coordinate	260,000	273,100	286,200	299,300	312,400	325,500	338,600	351,700	364,800	377,900	391,000	404,100	417,200
709,000	–	–	–	–	–	–	–	–	–	–	–	–	–
721,150	–	–	0.17	0.14	0.10	–	–	–	–	–	–	–	–
733,300	–	–	0.20	0.13	0.13	0.08	−0.01	−0.10	–	–	–	–	–
745,450	–	0.19	0.32	0.21	0.16	0.11	0.04	0.01	0.09	0.15	–	–	–
757,600	–	0.21	0.25	0.24	0.23	0.02	0.10	0.21	0.13	0.18	0.27	0.00	–
769,750	–	0.20	0.20	0.16	0.17	0.14	0.22	0.46	0.38	0.16	0.19	0.16	–
781,900	–	0.02	0.08	0.03	0.07	0.04	0.21	0.28	0.31	0.28	0.24	0.30	–
794,050	–	–	0.12	0.10	0.09	0.06	0.01	0.16	0.23	0.26	0.30	0.33	–
806,200	–	–	–	0.19	0.16	0.15	0.15	0.15	0.17	0.23	0.25	0.24	–
818,350	–	–	–	–	–	0.23	0.21	0.21	0.21	0.23	0.25	0.25	–
830,500	–	–	–	–	–	–	–	0.27	0.26	0.26	0.27	0.28	–
842,650	–	–	–	–	–	–	–	–	–	0.31	0.31	0.31	–
854,800												0.35	
866,950													

Table 13 Results of parameters' interpolation in AR models using 51 series by the nearest neighbor method

Coordinate	260,000	274,500	289,000	303,500	318,000	332,500	347,000	361,500	376,000	40,500	419,500	416,750	416,750
732,000	0.18	0.18	0.18	0.18	0.18	−0.14	−0.14	−0.14	−0.14	−0.14	0.14	0.14	0.14
741,300	0.18	0.18	0.10	0.10	0.10	0.17	−0.14	−0.14	−0.14	0.14	0.14	0.00	0.00
750,600	0.29	0.18	0.18	0.14	0.17	0.17	−0.14	−0.14	0.25	0.14	0.14	0.00	0.00
759,900	0.11	0.29	0.45	0.14	0.17	0.17	0.09	0.12	0.25	0.14	0.14	0.00	0.00
769,200	0.11	0.29	0.22	0.16	0.31	−0.08	0.11	0.12	0.25	0.32	−0.04	0.00	0.00
778,500	0.05	0.29	0.29	0.31	0.19	0.26	0.59	0.43	0.08	0.32	0.08	0.21	0.21
787,800	0.05	−0.02	0.08	0.15	0.00	0.28	0.28	0.22	0.37	0.27	0.43	0.21	0.21
797,100	0.13	0.13	−0.13	0.15	0.15	−0.13	0.10	0.23	0.23	0.31	0.43	0.16	0.16
806,400	0.13	0.13	−0.13	0.15	0.15	−0.13	0.10	0.19	0.19	0.31	0.16	0.16	0.16
815,700	0.13	0.13	−0.13	0.15	0.15	−0.13	0.19	0.19	0.19	0.31	0.16	0.16	0.16
825,000	0.13	0.13	−0.13	0.15	0.15	−0.13	0.19	0.19	0.19	0.19	0.16	0.16	0.16
834,300	0.13	0.13	−0.13	0.15	−0.13	0.19	0.19	0.19	0.19	0.39	0.39	0.39	0.39
843,600	0.13	0.13	−0.13	0.15	−0.13	0.19	0.19	0.19	0.39	0.39	0.39	0.39	0.39
852,900	0.13	0.13	0.15	0.15	−0.13	0.19	0.19	0.39	0.39	0.39	0.39	0.39	0.39

References

1. Akima, H.: A method of bivariate interpolation and smooth surface fitting for irregularly distributed data points. ACM Trans. Math. Softw. **4**(2), 148–159 (1978)
2. Akima, H.: Algorithm 761: scattered-data surface fitting that has the accuracy of a cubic polynomial. ACM Trans. Math. Softw. **22**, 362–371 (1996)
3. Bailey, T. C., Gatrell, A. C.: Interactive spatial data analysis. Addison Wesley Longman, Essex (1995)
4. Bărbulescu, A.: A new method for estimation the regional precipitation. Water Resour. Manag. **30**(1), 33–42 (2016). doi:10.1007/s11269-015-1152-2
5. Bărbulescu, A., Deguenon, J.: Nonparametric methods for fitting the precipitation variability applied to Dobrudja region. Int. J. Math. Models Methods Appl. Sci., **6**(4), 608–615 (2012)
6. Bivand, R., Pebezma, E. J., Gomez-Rubio, V.: Applied Statistical analysis with R. Springer, Berlin (2008)
7. Chang, C.L., Lo, S.L., Yu, S.L.: The parameter optimization in the inverse distance method by genetic algorithm for estimating precipitation. Environ. Monitor. Assess. **117**, 145–155 (2006)
8. Chen, D., Taki, R., Fan, W.: Using a genetic algorithm to optimize the inverse distance weight (IDW) interpolation method. In: Seventh International Conference on Geographic Information Science, Columbus, OH (2012)
9. Chiles, J.P., Delfiner, P.: Geostatistics, modeling spatial uncertainty. Wiley, New York (1999)
10. Cleveland, W.S., Devlin, S.J.: Locally weighted regression: an approach to regression analysis by local fitting. J. Am. Stat. Assoc. **83**(403), 596–610 (1988)
11. Cressie, N.A.C.: Statistics for spatial data. Wiley Series in Probability and Mathematical Statistics, New York (1993)
12. Davis, J.C.: Statistics and data analysis in geology. Wiley, New York (1986)
13. Isaaks, E.H., Srivastava, R.M.: Applied geostatistics. Oxford University Press, USA (2007)
14. Johnston, K., Ver Hoef, J.M., Krivoruchko, K., Lucas, N.: Using ArcGIS geostatistical analyst (2003). http://dusk.geo.orst.edu/gis/geostat_analyst.pdf
15. Lafitte, P.: Traité d'informatique géologique. Masson & Cie, Switzerland (1972)
16. Ledoux, H., Gold, C.: An efficient natural neighbour interpolation algorithm for geoscientific modelling. In: Fisher P.F (ed.), Developments in spatial data handling, pp. 97–108. Springer, Berlin (2005)
17. Li, J., Heap, A.: A review of spatial interpolation methods for environmental scientists. Geoscience Australia, record 2008/23, 137 pp
18. Li, J., Heap, A.D., Potter, A., Daniell, J.J.: Application of machine learning methods to spatial interpolation of environmental variables. Environ. Model Softw. **26**, 1647–1659 (2011)
19. Lu, G.Y., Wong, D.W.: An adaptive inverse-distance weighting spatial interpolation technique. Comput. Geosci. **34**, 1044–1055 (2008)
20. Ly, S., Charles, C., Degre, A.: Different methods for spatial interpolation of rainfall data for operational hydrology and hydrological modeling at watershed scale: a review. Biotechnologie, Agronomie, Société et Environnement **17**, 67–82 (2013)
21. Plush, P.: Kriging and splines: theoretical approach to linking spatial prediction methods. Interfacing Geostat. GIS, 45–56 (2009)
22. Pollock, D.S.G.: Smoothing with Cubic Splines, r.789695.n4.nabble.com/file/n905996/SPLINES.PDF
23. Sama, D.D.: Geostatistics with applications in earth sciences, 2nd edn. Springer, Berlin (2009)
24. Sanaei Nejad, S.H., Ghahraman, B., Pazhand, H.R.: Extended modified inverse distance method for interpolation rainfall. Int. J. Eng. Inventions **1**(3), 57–65 (2012)
25. Shepard, D.: A two-dimensional interpolation function for irregularly-spaced data. In: Proceedings of the 1968 ACM National Conference, pp. 517–524 (1968)
26. Sibson, R.: A brief description of natural neighbour interpolation. In: Barnett, V. (ed.) Interpreting multivariate data, p. 2135. Wiley, Chichester West Sussex, New York (1981)

27. Thiessen, A.J., Alter, J.C.: Precipitation averages for large areas. Mon. Weather Rev. **37**, 1082–1084 (1911)
28. Wahba, G., Wendelberger, J.: Some new mathematical methods for variational objective analysis using splines and cross-validation. Mon. Weather Rev. **108**, 1122–1145 (1980)

Printed in the United States
By Bookmasters